Microbes and Other Shamanic Beings

César E. Giraldo Herrera

Microbes and Other Shamanic Beings

palgrave
macmillan

César E. Giraldo Herrera
Somerville College
Institute for Science Innovation and Society
School of Anthropology and Museum Ethnography
Oxford, UK

ISBN 978-3-030-10041-4 ISBN 978-3-319-71318-2 (eBook)
https://doi.org/10.1007/978-3-319-71318-2

Softcover re-print of the Hardcover 1st edition 2018

Cover illustration: Carlos Jacanamijoy Navegación Interna (2009), oil on canvas

Printed on acid-free paper

This Palgrave Macmillan imprint is published by Springer Nature
The registered company is Springer International Publishing AG
The registered company address is: Gewerbestrasse 11, 6330 Cham, Switzerland

Para Enrique:
Mugre en mis ojos
manchas de tigre

Prologue

This book revaluates familiar myths and understandings of the world with insights developed while doing fieldwork far away from home. It is part of a trilogy derived from my doctoral dissertation in social anthropology. The dissertation dwelt on Nordic seamanship, on relating to the environment without and within, on syncretism, perspectivism, and shamanism. This book is the last part, the trip back home, back into ourselves. So, what or who are we?

My grandfather was a rural medic, in Quindio, Colombia a place infested with venomous snakes. He was adamant that one should get to know them, to see the world from their perspective, acknowledging their ecology, where they lived, and what they ate. Most of their attacks were in self-defence, out of fear. If in your interactions with them you were calm and respectful, even the most poisonous were mostly harmless. My grandfather supported his views on animal subjectivity with authors like Konrad Lorenz. He would have been an avid reader of Ingold, Haraway, Bennet, and Tsing. However, the roots of his views are more likely to be found in the adventures of the miscreant uncle Rabbit and his victimized predator uncle Jaguar, which he used to tell me. Although warranting care for deceiving appearances, these were more than fables, these are far older stories.

I studied my undergrad in biology, focusing on physiology, ethology, and theoretical biology, and so, I am an unrepentant functionalist and would not hesitate to subscribe to a naturalistic[1] understanding of reality.

However, my degree monograph was a theoretical exploration of biosemiotics, suggesting that organisms and other biological systems develop processes of interpretation, and in their own ways, and for their own sake make sense of their world. I find it very hard to fathom how some people assume humans are the only beings with intentions, points of view, or emotions: the only beings to communicate, the only persons. My views, as well as those of my parents, and grandparents, are in many ways closer to what some authors would denominate animism. So, again, who are we? We are the Westernized, or rather their descendants, and we are also descendants of the *indianos*, the Indianized, the Africanized; the colonizers, and the colonized.[2]

When I was little, I was terrified of the night and fascinated by those monsters that linger in Latin American imaginaries: the witches who transform into jaguars; the one-legged *Patasola;* the *Duende,* a goblin with a humongous hat and backward feet; and the *Llorona,* the spirit of the woman who cries for her abandoned baby. I felt more sympathy for the *Madremonte,* the mother of the forest, and for the *Mohan*, the mischievous guardian spirits of the waters. However, their stories, when they had them, were so diluted and abstract, that they had become caricatures, folkloric fictions.

Years later, working with traditional Afro-Colombian fisher peoples in the Pacific, I met with 'the visions', a rich oral tradition referring some of those stories, linking them with specific ecologies, medicinal plants, behaviours, and powers. The Tunda was a one-legged Amerindian woman (Afro-Colombian for Amerinds), who would appear in the shape of a close relative or a lover, and lure her victims into the wildness of the mangrove, taking away their speech, reducing them into sexual slavery and madness. However, there are also herbs and prayers to call *Tunda*; she teaches her protégés the art of invisibility, and hides them from the authorities. These visions could also appear and harass you in dreams. My friends and hosts interpreted some of my own dreams in that way. However, the visions seem to flee from modernity, disappearing together with the ecologies with which they are associated. The visions made evident that the folkloric monsters of my childhood were translations of the masters of game, some of the beings with which Amerindian shamans deal. I sought to explain them as symbolic constructs, enunciating the affordances and dangers of specific environments and the social relations people established in them.[3] However, these interpretations neglect the experiences associated with these beings, how people understand them, and the ways people seek to interact with them.

Later still, reflecting upon my experiences while on-board the industrial fishing trawlers in the North Sea, I began to explore how we relate to microbes, how we may perceive them, and came across a possible alternative translation, which would seem to account for more of the characteristics of masters of game. This led me to explore the early records of Amerindian shamanism and Amerindian myths associated with syphilis, developing a biocultural ethnohistory of Amerindian shamanism and microbiology.

Microbes and other shamanic beings explores whether and to what degree microbiology might be commensurable with shamanism, whether it might offer better translations than anthropology, following missionary theology has so far. The book develops three major arguments. First, shamanism has been generally understood through reference to spirits and souls. However, these terms were introduced by the missionaries, who carried the earliest translations, to convert Amerinds into Christianism. Rather than trying to comprehend shamanism through medieval European concepts, we should examine it through ideas that started developing in the West only after encountering Amerindian shamans. Since the earliest accounts, Amerindian shamanic notions have shared more in common with current microbial ecology than with Christian religious beliefs. Shamans have described the beings with which they deal in ways that correspond to contemporary understandings of microbes. Second, various human senses allow the unaided perception of the microbial world. We focus on entoptic vision, which affords the perception of microscopic objects flowing through our retina. The techniques employed by shamans enhance these kinds of perception, and their depictions of shamanic beings correspond to the images produced by these forms of vision. Third, the theory that some diseases are produced by living agents acquired through contagion was proposed near after the Encounter by a physician who translated and adapted Amerindian knowledge about syphilis, an important subject of pre-Contact Amerindian medicine and mythology. Amerindian myths of the Sun and the Moon described shamanic beings causing syphilis and closely related diseases, their dynamics, histories, and treatments. Western medicine took four centuries before revaluating its paradigms, rediscovering germs, and turning microbiology into a mainstream science. I argue that a deep decolonization of thought should reclaim this knowledge back. At a time when the war on microbes is becoming unsustainable, shamanism may afford a refined diplomacy to interact with the highly social microbial worlds which constitute and permeate us.

References

Descola, Philippe. 2013. *Beyond nature and culture*. Trans. J. Lloyd. Chicago: University of Chicago Press.

Giraldo Herrera, César Enrique. 2009. *Ecos en el arrullo del mar: Las artes de la marinería en el Pacífico colombiano y sy mimesis en la música y el baile, Prometeo.* Bogotá: Uniandes – Ceso – Departamento de Antropología.

Herrera Angel, Marta. 2016. *El conquistador conquistado. Awás, Cuayquer y Sindaguas en el Pacífico colombiano, siglos XVI–XVIII*. Bogotá: Ediciones Uniandes.

Viveiros de Castro, Eduardo. 1998. Cosmological deixis and Amerindian perspectivism. *Journal of the Royal Anthropological Institute* 4 (3): 469–488.

Acknowledgements

Being and thinking are collective processes, hence so are researching and writing. I have been fortunate to enjoy good company, insightful interactions, inspiring conversations, hospitality, and sponsorship throughout this adventure. Thus, I am deeply indebted to innumerable persons (human, non-human, and institutional) who have made this work possible, and although I will certainly miss many important people, I will try to evoke some of their names.

From the moment of his conception, Enrique, my son, has been a constant source of hope and inspiration, the most enlightening, exhilarating, nerving, and also frightening experience. With him I am beginning to understand my parents, Marta Herrera and César Giraldo, their caprices, their joys, and their unfaltering support. I have also counted with the hospitality and the constructive criticism of various texts of my aunts Leonor Herrera and Carmenza Charrier.

Through my doctoral research at the University of Aberdeen, I was fortunate to enjoy the guidance and supervision of Tim Ingold. The generous care he puts onto reading, commenting, and discussing ideas has allowed us, his many students, to experience the craftsmanship of knowledge. It was an honour and a delighting insightful experience. With his playful attitude, ruminated words, and gestures encompassing the world, I could set forth to explore reality at sea, in books, in my body, and imagination. Through the first steps of this research, my ex-wife, Angélica

Quintero, was a good listener who patiently supported me in my endeavours. I also had the chance of reading and being read by many colleagues, sharing good laughs of amazement with Caroline Gatt, Peter Loovers, Miriam Rabelo, Cristián Simonetti, Jennifer Clarke, Rachel Harkness, Nicolas Ellison, James Leach, Nancy Wachowich, Arnar Árnason, Robert Wishart, Ursula Witt, and many undergraduate students. I am also grateful to my examiners Kay Århem and Gísli Pálsson for their challenging and encouraging readings of the doctoral dissertation, for pointing at missing paths, and for suggesting alternative routes.

Later, Gísli Pálsson took me under his wing as a postdoctoral researcher for the project *Biosocial Relations and Hierarchies* at the University of Iceland, allowing me to explore the resonance of these ideas working on biosociality, developing research with microbiologists, and lecturing the course Body and Society. I also want to thank Guðmundur Hrafn Guðmundsson, and other members of his lab, as well as the researchers from Matís Prokaria, for sharing with me their insights on microbes. During this time, I presented a paper on shamanic microscopy (largely the Second Part) at the *Spaces of Attunement Symposium*; the comments of the participants as well as those from various anonymous reviewers were very helpful, and a version of this paper was accepted for publication at the *Journal Anthropology of Consciousness.*

During the past two years, Somerville College at the University of Oxford has nurtured me, providing an environment to grow and develop my ideas. I want to thank all its staff and students for their friendship, support, and perceptive conversations, in particular its former Principal Alice Prochaska, Maan Barua, David Bowe, Siddarth Arora, Alfred Gathorn-Hardy, and Philip Kreager. I also want to thank the staff and students of the Department of Anthropology, specially the members of the Institute of Science Innovation and Society, who welcomed me: Javier Lezaun for his tutorship, Steve Rayner, Jerome Ravetz, Christopher Goldsworthy, Louise Bezuidenhout, Sara de Wit, Rob Bellamy, Sophie, Hainess and Lisa Dilling for their insightful readings of manuscripts of this text. I have also enjoyed seminars, courses, and conversations with Elisabeth Hsu, Kate Fayers-Kerr, Maryam Aslany, Laura Rival, Ramón Sarró, Stanley Ulijaszek, Lola Martinez, Paola Esposito, and Elizabeth Ewart. I also want to thank Elizabeth for inviting me to the ISCA seminars to present another version of the paper on shamanic microscopy, where I received very valuable feedback. I also want to thank Erica Charters and the other members of the Centre for the History of Science, Medicine, and Technology for inviting me to present a paper on Amerindian treponemal mythologies, and providing most insightful comments. I also had the chance to present this paper at the panel 'living well together' of the 2016 EASA Conference in Milan. I am very grateful to Jamie Lorimer, Beth Greenhough, Javier Lezaun, and Cressida Jervis-Read for organizing the Oxford Interdisciplinary Microbiome Project, providing venues to enrich and share novel developments on microbes.

I am most grateful for many enduring friendships: bewitching perspectives shared with Beatriz Ángel, the support through parenthood of Santiago Jara, the musical fabulations through mountains and forests, with Camilo Giraldo, and with Juanita Delgado for her perceptive being and enchanting songs. I also want to thank Juan Camilo Niño, Elisa Bale, Santiago Paredes, Franzi Carranza, Marta Herrera, and all the members of *Taller Umbra* as well as Marianne Cardale, and the participants of the Coloquio Chibcha for their constructive criticism of the text.

As a deeper research project develops from the book, I want to thank the hospitality and openness of Fabio and Nelson Yabur, as well as Abadio Green of the Gunadule communities of Ibgigundiwala and Maggilagundiwala, the interest, care and fascinating stories of the benkhun Alejandro Moya of the Wounaan community at el Papayo, and of William Mozombite, Ingano taita resident in Leticia.

This research would not have been possible without the financial sponsorship provided for my studies by the studentships of the College of Arts and Social Sciences of the University of Aberdeen and the Overseas Research Scheme Scotland. I am most grateful to Victoria Maltby for the Junior Research Fellowship from Somerville College of the University of Oxford, which has allowed me to bring these ideas to the fore.

I also want to thank Antonia Waldorf and many other anonymous reviewers who have contributed with their doubts, suggestions, and references.

Last but not least, a thousand thanks to the reader—I hope you enjoy this trip.

Notes

1. For biologists, a naturalist is a researcher who after many years in the field has come to understand the characteristics of an ecological system, the organisms that make up its community, their behaviour, physiology, developmental, and genetic histories. It is the sort of thing you want to be when you grow very, very old. So, when I read how naturalism is being portrayed in anthropology (Viveiros de Castro 1998; Descola 2013), I cannot but feel profoundly annoyed by what seems like a strawman made with Christian hay and humanist clothes. However, that naturalism is a metonym, which names the whole, that is, that despair bundle of Western ontologies, by what today is one of its most prominent strands.
2. Herrera Angel 2016.
3. Giraldo Herrera 2009.

Contents

List of Figures

Check for updates

1

Colonized Selves, Decolonizing Ontologies

1.1 Decolonizing Ontologies

We can define ontology as the knowledge or understanding of being or reality. I prefer these terms over discourse, science, study, and many other alternatives, because in general these alternatives imply formalized approaches to reality. Understanding is a more inclusive, down-to-earth, dynamic notion, which captures the implications of ontology, its capacity to make worlds. Understanding and being are imbricated. Understanding a reality, we articulate what is known about it, what we perceive, and what we infer from those perceptions into something coherent, which we can act upon. On the other hand, quite literally understanding fundaments being, and becomes the basis for reality.

Like the God of Christians, reality, science, and ontology used to be employed solely in the singular and capitalized. While Christianity was thought to convey the 'true knowledge' of the 'one true God', so Science (singular and capitalized) was thought to be the rational, incremental process of acquiring knowledge: the understanding of the 'true' and 'universal' nature of Reality (again singular and capitalized)—a process originating in Europe, enabling the progress of the West and, implicitly or explicitly, justifying its colonization of the rest of the world.

C. E. Giraldo Herrera, *Microbes and Other Shamanic Beings*,
https://doi.org/10.1007/978-3-319-71318-2_1

Anthropology, as part of this enterprise, would recognize isolated achievements of non-Westerners, such as the botanical knowledge of some Amerindian shamans. Nevertheless, these were frequently downplayed as local or situated forms of knowledge and contrasted with the 'universal' character of Western Science. Moreover, the understandings of reality of shamans were frequently dismissed, because they included entities such as 'souls' and 'spirits', which modernity had banished from Nature or Reality. To make matters worse, shamans attributed these entities to non-humans, even to inert or inanimate things like rocks. As remarked by Descola (1996), these entities and their characteristics still remain ever-perplexing.

Over the past 50 years, anthropology has been conducting a profound critique of its position, its relation to its subjects of study, and its relations to power. Through this reflexive process, anthropology has come to recognize its own role in colonial and neo-colonial processes and to question its methods, and scientific pretentions, as well as the scientific enterprise and the process of development as such.

Post-colonial intellectuals have revealed that colonialism involves the authoritative deployment of artistic, literary, academic, and scientific discourses, including anthropological discourses of the Other, as forms of epistemic violence, undermining how non-Western peoples perceive and understand themselves and reality. This colonization of thought underlays the control, suppression, and exploitation of non-Westerners even after they become politically emancipated.[1] Real emancipation requires a decolonization of thought, a re-evaluation of non-Western forms of art, of telling stories, of thinking, and understanding the world.[2]

On the other hand, applying to natural scientists the methods anthropology developed to understand ethnoscience, science and technology studies (STS) have observed that the claims of natural scientists are substantiated, and they derive their strength from highly situated practices, articulating human and non-human actors through arduous processes of negotiated translation. These practices are possible under specific historical contingencies; they are embedded in social dynamics, such as the politics of the academy, and diverse conjunctures with finance, industry, and religion.[3] Moreover, when different disciplines address 'the same reality', they define and articulate it differently, often reaching dissimilar conclusions.[4] Science stands for a plurality of different practices, views, and voices, which are frequently in disagreement. Thus, the universality, and even the unity, of Science has been an idealistic aim rather than an actual achievement.

Examining the history of science in colonial and neocolonial settings, postcolonial historians of science have demonstrated how Western scientific models and projects of development are frequently founded on ethnocentric assumptions, which cannot be generalized to conditions far removed from the working parameters in which the models were developed, leading to scientific stagnation and disastrous developmental failures. Like any other system of knowledge, Western sciences have a limited grip on reality. Furthermore, distant allegiances frequently draw the interests of scientists and developers, aligning them with those of the Western metropolis, to the detriment of the 'peripheries,' and of knowledge itself. The recognition of these inadequacies and limitations resulted in the call for cognitive justice, for the recognition that there is a plurality of sciences, that other forms of knowledge may lay better claims to understanding the true nature of reality,[5] thereby opening another path for the decolonization of thought. Western sciences cannot be the yardstick to judge the validity of non-Western ontologies, that is, the way non-Western realities are understood and constituted.

1.2 The Ontological Turn and Its Challenges

In the past years, these insights have led anthropology into the ontological turn—the realization that our interlocutors not only have different cultures but also often dwell in radically different realities. Consequently, there are multiple ontologies, which might not be commensurable. Radical difference and incommensurability foster the study of ontological claims on their own terms, seeking to approach diverse realities and their understandings without privileging one (the Western) over the others. This bolder version of the classical relativistic principle of anthropology (which encourages us to understand social practices and behaviours within their cultural context) extends relativism onto reality itself. This should be regarded as a point of departure for an anthropology seeking to understand: what is being, what is the world, and how do we and our interlocutors know it?

As Holbraad[6] points out, the ontological turn places the onus of proof on the anthropologist, who has to 'reconceptualise a whole set of notions' to address whatever our interlocutors are dealing with. He indicates that we ought to give precedence to their understanding of reality over our interpretations of how they understand reality. If our concepts raise paradoxes, it is these concepts that we must re-evaluate. He suggests we should re-evaluate what things are (e.g., What is a stone? for it to have a soul or a spirit).

Following the epistemic principle of the ontological turn, Vilaça (2005) and Holbraad et al. (2014) have supposed that to decolonize thought, it would be sufficient to uphold current shamanic and scientific ontologies as equivalent. They assume that by simply acknowledging their incommensurabilities, ontologies that have become subaltern can be explained and defended in their own terms. Thus, they argue that a decolonization of thought should focus on non-biological conceptions of the body and the environment. Following this path, anthropology has enthusiastically assumed that non-Western realities, like those of shamanism, are necessarily incommensurable with those portrayed by natural sciences.

However, as is noted by Descola, biological and ethnobiological classifications and understandings most frequently coincide in their details.[7] Furthermore, as is suggested by Latour in his comments to Eduardo Kohn's *How Forests Think*, by itself, an extended relativistic principle in anthropological epistemology does not necessarily empower non-Western realities or their understandings:

> … how could an ethnographer, or, for that matter, a Runa scholar, equipped with such a philosophical anthropology *find ways to make his or her ontological claims understood in negotiating what a forest is* made of, when faced with forestry engineers, loggers, tourists, NGOs, or state administrators? *That is where the so-called ontological turn finds its moment of truth. Not on the epistemological scene but on the bittersweet attempts at negotiating alternative ways to occupy a territory*, being thrown in the world, designating who is friend and who is enemy.[8] (Emphasis is mine)

Western and non-Western peoples increasingly coexist in the same places, partially sharing common realities. Even if these realities are articulated into distinct worlds and ontologies, they overlap and become subjects of dispute. Shamans and biomedical practitioners treating Amerindian or other communities share the realities of health and disease in these populations. Because of these common grounds, they often end up competing or interfering with one another, potentially to the detriment of their patients.

In those disputes, a powerful side might seek to settle the issue unilaterally. However, such resolutions are often resisted, or contested.[9] Authority and influence, also depend on the degree of specific understanding of the incumbent realities, and the capacity to make this knowledge operational. For example, the knowledge about particular diseases, the ability to treat them, the availability of therapeutic means or apparatuses, and the capacity to communicate knowledgeability and mobilize patients may ultimately grant authority to either shamans or biomedical doctors. Authority also depends on the capacity to translate operational understandings to a language acknowledged by the counterpart. Although shamans have made multiple contributions to Western pharmaceutics, the metaphysical or supernatural terms in which their knowledge is translated lead biomedical practitioners to continue to dismiss their practices as fake and superstitious.

I will argue that although questioning the reality of the stone is bold, the stone should not be the main source of concern for the ontological turn. After all, Amerindian shamans, anthropologists, and natural scientists can normally perceive rocks and acknowledge them as real, and therefore at least as partially commensurable. What is really troubling is: what is a soul or a spirit, for it to be in a stone? Because, strictly speaking, the Christian religious notions of souls and spirits are incommensurable with naturalist understandings of both stones and humans.

Moreover, as Davi Kopenawa, a Yanomami shaman, points out, it is anthropologists who translate the Yanomami *xapiri* as spirits,[10] and, as has been recognized since Lévy-Bruhl and is still quite evident,[11] the translation is not very good. The Christian religious notions of souls and spirits are incommensurable with naturalist understandings of stones and humans, but also with shamanic notions of 'spirits' and 'souls'. What is problematic is the continuous translation of Amerindian notions through Christian religious terms, the subsequent reduction of Amerindian realities onto the realm of the supernatural, and of their ontologies onto

Christian metaphysics, to such a degree that shamanism is reduced into the 'belief in the existence and manipulation of spirits', and ultimately it is assumed to be a matter of faith not of knowledge. What is troubling are the explanations built upon these translations and delivered by the anthropology of shamanism, not the explicative capacities of shamanism itself. This translation is a classic example of epistemic violence. Anthropology needs to address critically its theological missionary roots, how Christian categories affect its understanding of reality, and its relations with the knowledge of its interlocutors.[12]

Anchored in theological language, anthropology has found itself to address the corporeal and organismic characteristics of beings like *xapiri*. How could these beings have their abode in human and non-human bodies, in objects and in artefacts? Anthropology has also faced difficulties in understanding and acknowledging their pathogenicity; how could these beings cause diseases? Specially, how could they cause infectious–contagious diseases? But also, how could they affect pregnancy, birth, or development? Furthermore, so far, anthropology has lacked the appropriate conceptual framework to consider seriously the causal relations shamanism establishes between the management of wildlife and that of health.

Praet (2009), who also notes the dilemma of translation, offers 'monsters' as an alternative to spirits, gods, deities, and other similar terms. He provides an insightful etymology of the term. Nevertheless, this translation aggravates the problem. If the existence of spirits and gods is a matter of faith, monsters are generally not believed in. On the other hand, this translation also fails to account for the crucial characteristics of these beings, mentioned above.

Perhaps we should leave notions like the Yanomami *xapiri* untranslated. But do they ever remain untranslated? Evidently not. Christian terms and their universalist interpretations continue to creep into ethnographies and theoretical debates. On the other hand, if we could translate them, why should these notions remain untranslated while jaguars and trees are not.

However, these notions might have been translated already. As is pointed out by Raj (2013), the call for cognitive justice, the recognition of the situated character of science, and I would add the ontological turn have assumed that different ontologies or forms of knowledge have developed in isolation and somehow remained hermetically refractory, neglecting that a crucial locus for the development of knowledge is its circulation—not its simple diffusion or its faithful transmission, but the negotiated process of its multiple reinventions.

Since at least the sixteenth-century, Amerindian, African and European cultures have undergone intensive processes of hybridization or syncretism.[13] At various times, Amerindian knowledge may have been influenced by Western explorers, missionaries, and scientists, or by the policies devised after their work.[14] However, these scientists and explorers were immersed in non-Western societies and were more likely to be influenced by their idiosyncrasies and ontologies.[15] They drew data, inspiration, and theories from extra-European sources, which frequently remained unacknowledged.[16] Anthropology and history are only beginning to explore how syncretism has constituted the West, to examine the global histories of science.[17] Although this research is a promising start, most of the work exploring syncretism between the New and the Old Worlds has concentrated on the Baroque or Enlightenment.[18] The temporal bias of the literature limits the possibility of assessing the true effects of the Encounter in the constitution of the West. Moreover, these late dates promote the deceitful impression that Europeans always ascribed to naturalism, and gave precedence to scientific thought over other forms of understanding the world.

Anthropology, history of science, and STS have largely neglected that natural history and natural philosophy were not the dominant discourses in Europe before the Encounter. Through the Middle Ages and the Renaissance, natural philosophy was at best a handmaid for theology. The early European ambassadors of knowledge were missionary friars, who believed in the sacredness of religious icons, and credited faith above empirical or objective knowledge. This calls into question depictions of the encounter of West–non-West that assume a naturalist or objectivist stance on behalf of the Europeans.[19] Nevertheless, those same missionaries carefully collected Amerindian natural knowledge. In this way, missionaries and colonial officials participated in processes of *purification* and *translation*,[20] whereby Amerindian knowledge was decontextualized, differentiated into moral and natural histories, adapted, repurposed, and appropriated.

In this light, the decolonization of thought proposed by advocates of the ontological turn seems unsatisfactory because it assumes that natural sciences are indeed monolithic. It neglects the appropriation and influence of non-Western knowledge in the constitution of Western sciences and, what might be worse, subsumes Amerindian ontologies to those of European theology and humanism. I contend that decolonizing thought should also involve reclaiming it, recognizing how natural sciences are also rooted beyond the West and how they can also allow us to explore and to acknowledge the weight of non-Western ontologies, affording for closer-fitting translations, and stronger alliances with entities that have biopolitical leverage in the global scene.

This book argues that Amerindian shamanism might be better understood through notions that arose in the West only after Contact, such as the theory of contagion. The descriptions of shamanic beings match closely and may be commensurable with currently developing understandings of microbes. While the relations shamans propose fall nearer to the purview of ecology and medicine, microbiology might facilitate a better *translation*[21] than the one the anthropology of shamanism has provided so far.

Anthropology has often regarded microbes with the same critical disbelief awarded to the 'spirits' described to ethnographers by animist hunters and shamans. Thus, when microbiologists raise concerns about the catastrophic consequences of our war on germs, or the rise of emerging diseases, these critical issues seem far removed from the reach and the interest of anthropology. Only recently, have researchers in STS begun to develop an understanding of how microbes and microbiologists mutually constitute one another and crucial dimensions of our everyday reality.[22]

Nonetheless, STS and anthropology have largely based their insights on the assumption that instruments forcibly mediate human perception of microbes.[23] Continuing the traditional views of the history of science, microbiology is generally assumed to be a science without precedents and a reality incommensurable with those from the past and with non-Western medical traditions, such as shamanism. The estrangement of these worlds forecloses the possibility of understanding how these health and environment management traditions have contributed to the constitution of contemporary microbes and of microbiology.

1.3 Commensurability, Translation, and Deep Decolonization: Reclaiming Knowledge

My overall intention is to develop a better translation, a translation that transcends Western popular knowledge of natural sciences and, as Holbraad proposes, acknowledges the degree of expertise involved in specialized trades like those of shamanic practitioners. Such translation should relate each others concepts but also they should articulate relations with beings in the world, establishing alliances that empower non-Western ontologies.

The validity and weight of a translation depend on the commensurability of the terms involved. This commensurability should be demonstrated to the counterpart (i.e. shamans and microbiologists). The demonstration of commensurability must provide evidence that there are equivalent epistemologies, granting access to the same realities. That is, the different understandings of reality are derived from experiences that can substantiate similar insights, or that there are modes of reasoning that can grant access to the same conclusions. To ascertain whether microbiology is a suitable fit for shamanism, we should demonstrate that microbes can be found in the circumstances where shamans locate shamanic beings and that they have similar characteristics. In the same way, to determine whether some shamanic beings might correspond to microbes or to a construction involving microbes, a crucial step is to examine how shamans may come to experience the microbial world or infer its existence and workings.

Translation can be an important aspect empowering non-Western ontologies. However, to proceed a step further in the decolonization of thought, we must reverse the process of purification to which non-Western ontologies have been subject, and which contributed to the emergence and growth of Western science. We should examine the genealogies of the ideas and practices of the fields of knowledge under comparison. Then, we should trace and question the genealogies of their Western counterparts, to examine whether they were in contact, whether their histories retain traces of syncretism, and whether their histories can illuminate our understanding of the history of non-Western ontologies. This allows us to recognize continuities and transformations in their modes of thought, to note the correlations in their development, but most importantly it expands their realm of meaning and serendipitously connects them with other realms of knowledge.

In a way, I am employing Latour's methods to question the historiography he proposes, indicating a deeper history of microbes that transcends the narratives of the history of Western science. I am inverting the double movement of *purification* and *translation*, re-constituting *factish* from fetish and facts.[24]

1.4 Description of the Book

To explore the avenues through which shamanism and microbiology might develop a productive conversation, this book develops a biocultural ethnohistory of shamanism and microbiology, through three parts. The **First Part** analyzes shamanism, exploring sixteenth- and seventeenth-century records of zemeism, shamanic knowledge, and practices from the circum-Caribbean area, showing the parallels of these understandings with the current understandings developed by microbiology. The **Second Part** examines means of perceptual access to the microbial world that are available to shamans and are consistent with their descriptions and depictions of entities like masters of game. The **Third Part** traces 16th and 17th records of Amerindian medical knowledge and a widespread Amerindian myth related to syphilis and other treponemal diseases, the syphilis pandemic of the sixteenth century, and the reception and adaptation of Amerindian knowledge by European medicine.

The **First Part** explores how Amerindian shamanism was framed, what it was, and alternative translations. First, Chapter 2 examines how anthropology understood, and misunderstood, shamanism and Amerindian ontologies through a conceptual framework derived from Christian missionary activities, specifically through the classical category of animism. Then, this chapter examines how, in the last 50 years, some anthropologists have reframed our understanding of animic thought. It explores the further extension of animic thought involved in Amerindian perspectivism and discusses and makes some amendments to this theoretical framework. Chapter 3 develops an ethnohistory of the shamanism of the Taíno and Callinago peoples of the Caribbean in their early contacts with Europeans. It examines European records of Amerindian understandings of the body, the environment, and the *zemes* or *çemijn* and other entities through which missionaries sought to translate their notions of gods, devils, or spirits. This chapter analyzes these ontologies in their own terms, as well as the practices and remedies employed by *buhities* or *boyáicou,* the Caribbean shamans, whom these authors described primarily as physicians, often noting their medical accomplishments, while simultaneously classifying them as magicians, charlatans, and adorers of the devil. Having provided an analysis of early Amerindian shamanism on its own terms, Chapter 4 explores the possibility of translating these notions through contemporary notions of microbiology. The chapter starts tracing possible continuities of Amerindian ontologies, comparing the ethnohistorical ontologies with those described by more recent ethnographies. It then examines the parallels between Amerindian shamanism and microbiology, demonstrating how microbiology is beginning to describe the world in ways that correspond to Amerindian shamanism. Finally, it addresses the objections to what could be seen as medical materialism and to the historical contingency of microbes.

Having shown how the recent developments of microbiology correspond to Amerindian shamanic ontologies, the **Second Part** explores the epistemology of Amerindian shamanism. Departing from the seventeenth-century records of an Amerindian inclination towards empiricism, we examine what means shamans have at their disposal for perceiving the microbial world, and show that their depictions of shamanic beings correspond to those perceptual means. The microbial world has been constantly before our eyes and under our noses. Microbes are perceivable on our skin and guts, tightly associated with our emotions, but can also be visible to the naked eye. We explore how microscopic entities are visible as they flow through our retinas, how their images can intermingle with those of our macro-reality, of our dreams and fantasies. Shamanic practices involve sophisticated techniques that optimize these forms of perception of the microbial world. First, Chapter 5 explores shamanic epistemologies, addressing the ways in which shamanic beings are not only conceptualized but perceived. Chapter 6 questions neuropsychological interpretations which undermine these experiences, explaining them exclusively as neurogenic hallucinations. In Chapter 7, we explore an alternative interpretation of shamanic vision: as a form of microscopy afforded by the structure of the retina of humans and other mammals.

We start exploring the anatomy and the physiology of the eye. Then we trace the history of research on entoptic phenomena and its different interpretations. Chapter 8 analyzes entoptic microscopy, what characterizes the images produced by these phenomena, whether these images of the microscopic world correspond to the descriptions and depictions of shamanic visions, and what can be seen through this form of vision.

The **Third Part** explores how Amerindian and European realities began to converge with the Encounter, and how sixteenth-century missionaries, explorers, and physicians translated Amerindian shamanic ontologies, adapting their naturalistic knowledge into European medicine, grounding the earliest *theory of contagion*, a scaffold for the development of microbiology. Chapter 9 analyzes Pané's records of a Taíno myth of Origin: a version of the Pan-Amerindian legend of the Sun and the Moon, in which syphilis and other treponemal diseases play a crucial role. These myths describe diseases, those afflicted by them, their pathogens, and some of their treatments as key players in the conformation of Taíno society. Chapter 10 examines a Nahuatl version of the same story, recorded by Bernardino de Sahagun, which likewise identifies these diseases as the Sun regent of Nahuatl peoples. Sahagun also records descriptions, classifications, and treatments of these diseases, which aid in the interpretation of the myth. Chapter 11 examines how Europeans came to share the realities of these diseases, traces how Amerindian shamanic ontologies and myths were translated and adapted by European physicians, influencing the formulation of the earliest theory of contagion in the West, constituting an ontological scaffold for the development of *Germ Theory* and ultimately microbiology. We examine current understandings of these treponemal diseases, showing how it is closer to sixteenth-century Amerindian ontologies.

Chapter 12 winds together the insights from the book. We summarize pre-Contact Amerindian shamanic ontologies describing zemes and other shamanic beings, and their specific instantiation in the case of *Guaynara* or *Nanahuatl*, the shamanic beings of treponematoses. We examine how and why missionaries and colonial officers translated Amerindian notions. We review how anthropological theory continued to replicate these interpretations. We review shamanic epistemologies, and the interpretation of visions as partly derived from entoptic microscopy, other forms perception and communication with the microbial world. After that we examine how shamanic beings like zemes were translated into semina, germs, and microbes. We conclude with a discussion about the decolonization of thought, and of the remaining incommensurabilities between microbiology and shamanism.

Notes

1. Spivak 1988.
2. Césaire 2006 [1955].
3. Latour 1987, 1988b, 1999.
4. Mol 2002.
5. Visvanathan 1998, 2006.
6. Alberti et al. 2011.
7. Descola 1996.
8. Latour 2014.
9. Jokic 2015.
10. Kopenawa and Albert 2013, 11.
11. Willerslev 2011.
12. Douglas 1984.
13. Gruzinski 2013; Sallmann and Gruzinski 1992.
14. Gudeman et al. 1989.
15. Wiener 2013; Latour 2010.
16. Safier 2008.
17. Sivasundaram 2010.
18. Bailey 2001, 2005.
19. e.g. Latour 2010.
20. Latour 2010.
21. *Translation* has two meanings here: the traditional idea of transferring meanings from one language to another, and the Latourian usage, articulating or forging alliances between different actants.
22. Latour 1988a; Paxson 2008; Paxson and Helmreich 2013; Tsing 2014; Helmreich 2009; Hird 2009.
23. Helmreich and Paxson recognize that microbes are pseudo-perceptible. However, this does not lead them to question the rhetoric of the history of microbiology.
24. Latour 1993, 2010.

Bibliography

Alberti, Benjamin, Severin Fowles, Martin Holbraad, Yvonne Marshall, and Christopher Witmore. 2011. "Worlds otherwise": Archaeology, anthropology, and ontological difference. *Current Anthropology* 52 (6): 896–912. https://doi.org/10.1086/662027.

Bailey, G.A. 2001. *Art on the Jesuit missions in Asia and Latin America, 1542–1773*. Toronto: University of Toronto.

———. 2005. *Art of colonial Latin America*. London: Phaidon.

Césaire, Aimé. 2006 [1955]. *Discursos sobre el colonialismo*, vol. 39. Madrid: Ediciones Akal.

Descola, Philippe. 1996. Constructing natures: Symbolic ecology and social practice. In *Nature and society: Anthropological perspectives*, ed. Philippe Descola and Gisli Palsson, 82–101. London: Routledge.

Douglas, Mary. 1984. *Purity & danger : An analysis of the concepts of pollution & taboo*. London: Routledge.

Gruzinski, Serge. 2013. *The mestizo mind: The intellectual dynamics of colonization and globalization*. New York: Routledge.

Gudeman, Stephen, Alberto Rivera, Richard Newbold Adams, Phyllis Deane, Jane I. Guyer, and Lynn Hirschkind. 1989. Colombian conversations: The strength of the earth [and comments and replies]. *Current Anthropology* 30 (3): 267–281.

Helmreich, Stefan. 2009. *Alien ocean: Anthropological voyages in microbial seas*. Berkeley: University of California Press.

Hird, Myra J. 2009. *The origins of sociable life evolution after science studies*. New York: Palgrave Macmillan.

Holbraad, Martin, Morten Axel Pedersen, and Eduardo Viveiros de Castro. 2014. The politics of ontology: Anthropological positions. *Cultural Anthropology Online*.

Jokic, Zeljko. 2015. *The living ancestors: Shamanism, cosmos and cultural change among the Yanomami of the upper Orinoco*. Oxford: Berghahn.

Kopenawa, Davi, and Bruce Albert. 2013. *The falling sky: Words of a Yanomami Shaman*. Trans. Nicholas Elliott and Alison Dundy. Cambridge, MA: Harvard University Press.

Latour, Bruno. 1987. *Science in action: How to follow scientists and engineers through society*. Cambridge, MA: Harvard University Press.

———. 1988a. *The Pasteurization of France followed by irreductions*. Trans. A. Sheridan and I. Law. Cambridge, MA: Harvard University Press. Original edition, 1984.

———. 1988b. The politics of explanation: An alternative. *Knowledge and Reflexivity: New Frontiers in the Sociology of Knowledge* 10: 155–176.
———. 1993. *We have never been modern*. Hertfordshire: Harvester Wheatsheaf.
———. 1999. *Pandora's hope: Essays on the reality of science studies*. Cambridge, MA: Harvard University Press.
———. 2010. *On the modern cult of the factish gods*. Trans. Barbara Smith and E Weintraub. Durham: Duke University Press.
———. 2014. On selves, forms, and forces. *HAU: Journal of Ethnographic Theory* 4 (2): 261–266.
Mol, Annemarie. 2002. *The body multiple: Ontology in medical practice*. Durham: Duke University Press.
Paxson, Heather. 2008. Post-pasteurian cultures: The microbiopolitics of raw-milk cheese in the United States. *Cultural Anthropology* 23 (1): 15–47.
Paxson, Heather, and Stefan Helmreich. 2013. The perils and promises of microbial abundance: Novel natures and model ecosystems, from artisanal cheese to alien seas. *Social Studies of Science*: 0306312713505003.
Praet, Istvan. 2009. Shamanism and ritual in South America: An inquiry into Amerindian shape-shifting. *Journal of the Royal Anthropological Institute* 15 (4): 737–754.
Raj, Kapil. 2013. Beyond postcolonialism… and postpositivism: Circulation and the global history of science. *Isis* 104 (2): 337–347.
Safier, Neil. 2008. *Measuring the new world: Enlightenment science and South America*. Chicago: University of Chicago Press.
Sallmann, Jean-Michel, and Serge Gruzinski. 1992. *Visions indiennes, visions baroques: les métissages de l'inconscient*. Presses Universitaires de France-PUF.
Sivasundaram, Sujit. 2010. Focus: Global histories of science. Introduction. *Isis; An International Review Devoted to the History of Science and Its Cultural Influences* 101 (1): 95–97.
Spivak, Gayatri Chakravorty. 1988. Can the subaltern speak? In *Can the subaltern speak? Reflections on the history of an idea*, ed. Rosalind Morris, 21–78. New York: Columbia University Press.
Tsing, Anna Lowenhaupt. 2014. Strathern beyond the Human: Testimony of a spore. *Theory, Culture & Society* 31 (2–3): 221–241. https://doi.org/10.1177/0263276413509114.
Vilaça, Aparecida. 2005. Chronically unstable bodies: Reflections on Amazonian corporalities. *Journal of the Royal Anthropological Institute* 11 (3): 445–464. https://doi.org/10.1111/j.1467-9655.2005.00245.x.
Visvanathan, Shiv. 1998. A celebration of difference: Science and democracy in India. *Science* 280 (5360): 42–43.

———. 2006. Alternative science. *Theory, Culture & Society* 23 (2–3): 164–169. https://doi.org/10.1177/0263276406023002 26.
Wiener, Margaret J. 2013. Magic,(colonial) science and science studies. *Social Anthropology* 21 (4): 492–509.
Willerslev, Rane. 2011. Frazer strikes back from the armchair: A new search for the animist soul. *Journal of the Royal Anthropological Institute* 17 (3): 504–526.

Part I

Amerindian Shamanism

Anthropology is filled with alluring subjects, with phenomena which do not translate easily, with misunderstandings that become definitions and allow us to embrace further unknowns. However, occasionally, it is a good idea to turn back and examine these old categories, to question the assumptions on which they were based—those assumptions that lent researchers some grip on worlds beyond their comprehension, to check whether there are alternatives, which may lend us a better grip. This part will explore what Amerindian shamanism is, what it was, and whether its claims can be partially commensurable with ontologies acknowledged by biomedicine and ecology. First, it will question how anthropology traditionally framed animism and the implications and impacts of these understandings for shamanism. After that, it will explore the novel ways of addressing animic thought. Then we will examine some of the earliest encounters with Amerindian 'shamans', the descriptions of their understandings of the world and their practices. We will then explore the commensurable features in the descriptions of microbes and shamanic beings. Finally, we will address the traditional resistance of anthropology to explore 'medical materialism' and question the historiographies of science which circumscribe microbes to the development of Western microbiology, with its instruments and methods.

2

(Mis)Understanding Shamanism and Animism

This chapter examines the notions of shamanism and animism. It questions their classical formulations, showing how they were based on conceptual frameworks inherited from missionaries. We proceed to examine how the notion of animism has been reformulated and revaluated, and then we will examine the *Theory of Amerindian Perspectivism.*

2.1 Entrenched Notions of Shamanism and Animism

Since the eighteenth century, when German explorers first described the charismatic Siberian shamans, the latter have fascinated Western imagination. The description of their paraphernalia, the ecstatic states they achieved through the use of psychotropic substances and the performance of singing, drumming, and dancing enthralled the Europeans. The narratives of their confrontations with 'spirits' and their flights through the skies to recall the lost 'souls' of ill persons captivated German philosophers like Kant and Goethe, who popularized the notion.[1]

C. E. Giraldo Herrera, *Microbes and Other Shamanic Beings*,
https://doi.org/10.1007/978-3-319-71318-2_2

Shamanism came to be described as a set of magic–religious practices involving the interaction with spirits through ecstasy or dreams, for the purpose of healing or divination.[2] The notion of shamanism came to designate a wide variety of practices of peoples around the Circumpolar North, Asia, and the Americas—peoples who were classified as animists during the late nineteenth century.

However, even within its original Siberian context, shamanic paraphernalia—the means shamans employ to acquire knowledge—along with the claims shamans make and their roles in society vary substantially. Some researchers have argued that the notion cannot be employed beyond its original contexts; others further claimed the animistic notions of spirits are the sole unifying characteristic behind the notion of shamanism.[3]

Tylor defined animism as the most basic form of religion, a crude philosophic model of nature involving the belief in spiritual beings that animate the world, providing plausible ultimate causation for the phenomena observed.[4] Animism represented 'the very essence of Spiritualistic as opposed to Materialistic philosophy'.[5] The basic model for these spiritual beings was a conception of the soul for the 'primitives', defined as following:

> It is a thin unsubstantial human image, in its nature a sort of vapour, film or shadow; the cause of life and thought in the individual it animates; independently possessing the personal consciousness and volition of its corporal owner, past or present; capable of leaving the body far behind, to flash swiftly from place to place; mostly unpalpable and invisible, yet also manifesting physical power, and especially appear to men after the death of that body; able to enter into, possess, and act in the bodies of other men, of animals and even of things.[6]

Tylor further remarks that this understanding of the soul persisted amongst the 'civilized', although it was modified and questioned by classical, medieval, and in particular modern philosophy.

The notion of the soul, he asserted, resolved for the 'primitive' as for the child, two different questions: what makes living a being? And, what are visions and dreams? He proposed that the duality of the dreaming condition, where one remains nearly inert and nonetheless experiences being active, suggested to the 'primitives' a plane of spiritual existence parallel to physical reality. The similarity between sleep and death helped 'primitives' explain what makes living a being, which would be the same soul, which is elsewhere while dreaming. The notion of the soul and the plane of spiritual existence, he claims, was further extended through the attribution of a soul-like spirit to non-human and frequently to 'inanimate' entities, providing an explanation for disease and other phenomena 'beyond the control' of animists.

Tylor's views were in line with Christian and evolutionist notions of the late nineteenth century. He characterized animism as a basic religion, distinguishing it from the monotheistic religion of civilized society, with its universal Deity ruling over Nature which, for him, was a precondition for the development of scientific thought. His explanations established the assumption of animism as the result of a failed or at least crude epistemology, infantilizing 'primitives'.

Despite its shortcomings, Tylor's account is not void of sophistication; it summons characteristic features that arguably can still be recognized articulated in the ontologies of multiple 'animist' cultures, which to this day continue to perplex anthropologists.[7] In Western terms, Tylor provided a parsimonious explanation for the connection of these features. Although theology has been excluded from scientific conversation and unilineal social Darwinism has been refuted, Tylor's understanding of animism has largely prevailed through the twentieth century in anthropological[8] and psychological[9] theory, as well as in the popular imagination. These theories have continued to regard animism as an early form of religion—the result of a naïve epistemology, which failed to grasp the 'universal and deterministic laws of nature'.[10]

The association of animism with religion and idealism forced onto animists Western categories that are critically regarded by science in the West, often by social researchers themselves.[11] As a consecuence of this, animists, and shamans in particular, have been considered as irredeemable antagonists to materialist science. Materialists systematically avoid dealing with animistic beliefs, even when trying to understand indigenous theories of disease or contagion.[12]

Because it is assumed that animism is based on a failed epistemology, shamanic practices have been likewise dismissed as delusion or self-delusion.[13] At best, it has been assumed that shamans operated through sophisticated metaphoric constructs representing human society, psychoanalysing their patients.[14] Either way, it is assumed that if shamans cure, it is only through the symbolic efficacy of placebo effects, a highly paradoxical turn considering that in the meantime the pharmaceutical industry has been profiting from the botanical knowledge that underlies many of their treatments.

2.2 Reframing Animism

Deterministic understandings of the physical world are being questioned and revised. The laws of physics are coming to be considered as exceptional cases,[15] isolated with great ingenuity from the actual complexity of the world.[16] Philosophy and social sciences have begun to question human exceptionalism, the idea that there is an incommensurable difference between humans and the rest of beings, a notion tightly linked to the mind–body dualism, which Western culture inherited from Christianism.[17] Humanities and social sciences have begun to explore how knowledge of the natural world is socially embedded, and how complex interactions between humans and other-than-humans constitute reality.[18]

Simultaneously, some anthropologists have begun to question and revaluate the understanding of animism. Descola brought back the notion after noting that the categories of culture and nature are far from universal. However, he remains very much in line with Tylor claiming that there is a universal assumption of a dualism between physicality and what he denominates interiority. The latter, Descola roughly defines as a set of properties, such as intentionality, subjectivity, reflexivity, and the ability to express oneself and to dream, which are recognized in all humans and encompassed by the mind or soul. He defines animism as the attribution of such an interiority to non-humans. Like Tylor, Descola is not void of sophistication—his definition of the qualities animists acknowledge to non-humans is precise. However, the universality of the dualism is questionable; as Descola himself notes, many of the terms that denote interiority refer to material and corporeal substances like blood, breath, fat, and latex.

Other researchers meanwhile have begun to recognize that animism is distant from the platonic assumptions of a simple nature that tends towards a static ideal, and requires an extraneous living force, a soul, or a mind to animate the inanimate. On the contrary, animists like the Ojibwa or the Gunadule recognize that everything in the world is intrinsically highly dynamic, changes, moves, and decays. Everything, even corpses for that matter, is subject to the processes of transformation. Hence, they acknowledge everything as animate,[19] and find the Western notion of the 'inert' or the 'inanimate' highly unrealistic and problematic.[20] Consequently, various Amerindian languages lacked a notion for the soul, or an all-encompassing category for animals, setting them apart from plants or other beings belonging to the same environments.[21]

Hallowell (2002 [1960]) observed that instead of classifying things as animate or inanimate, the Ojibwa acknowledge some rocks, plants, animals, technologies, and other beings as personae, which are characterized by their capacity to 'metamorphose', displaying extraordinary behaviours. These beings are liable to be affected but react in unforeseeable ways, and may be more powerful than could be assumed at first. They can have secondary or secret intentions, which require acknowledgement and a suitable treatment.[22] This, however, does not imply that animists fail to understand the mechanistic and 'orderly motion of things on the world', as psychologists assume.[23] These personae are not naively presumed of unknown things, as children would do. They are recognized amidst the deceptiveness of appearances, by experienced people dealing individually with known things, through deep and careful engagements, which may extend into dreams.[24] Through the course of such engagements, they come to recognize that the actual behaviour of the individual 'mechanism' is far from the idealized mechanization, may be far more complex. Thus, Ingold (2006) points out that rather than naively attributing a mind, a soul, or a live force to animate the inanimate, animic ontologies anchored in experience acknowledge the dynamic complexity of beings in their environment.[25]

As Descola (2013, 5) observed, dealing with individual beings often requires developing an intersubjective relation, which is mediated by techniques, but also by symbols and other signs, which often require acknowledging and taking into consideration their subjective point of view, the way in which the world would be seen from their perspective. Hence, animist thought has what Århem (1993) distinguished as a *perspectival quality*.

2.3 Amerindian Perspectivism

In Amerindian thought, a subjective persona may be acknowledged to individual beings behaving in extraordinary ways. However, many Amerindian communities also acknowledge such subjective perspectives to all the members of species of ecological or symbolic importance, or to masters (mothers) of game—entities which protect a given species counter-predating on its predators through disease. Moreover, Amerindian communities will distinguish other species of beings with their auto-ethnonym, the word they employ to designate their own ethnicity, a word which is often translated as *human, person,* or *people*. Seemingly paradoxically, these communities deny this status of 'humanity' to members of other ethnic groups.

Viveiros de Castro (1998) seeks to unravel this paradox, noting that these terms have a pronominal character—they are more similar to the pronoun *us,* rather than proper names. He argues further that unlike the Western evolutionary mythology in which humans become differentiated from animals, in Amerindian myths, animals once had human bodies, but through some moral failure, they diverged to their present forms. The differences between the given species of beings, be they humans, fish, jaguars, or shamanic beings, arise not on the ground of metaphysical or spiritual essences, but on account of the physical bodies, with their specific *weapons,* the instrument-organs through which they engage the world and the subsequent perspectives these weapons entail.

Because the body and its organs are not inert, but rather instruments with affects and affordances, shamans can employ these weapons, namely, claws, paws, animal skin, bile, blood, and multiple other substances, to perform *perspectival exchanges*, through which they transgress the limitations of the perspectives tied to their bodies. Such metamorphoses allow them to understand that each of these species perceives itself as the humans of those particular ethnicities perceive themselves, as the centre and measure of their world, as the bearers of technologies and cultural practices. The different species interpret the world through the parameters of a shared culture: customs, technologies, hierarchical organization, rituals, and intersubjective ecological relations with other species in the environment. Each perceives its sources of nourishment and its predators like humans perceive theirs: the jaguar-people perceive human-meat as manioc, which they cook, and the human-blood as manioc beer, which they brew. Fish-people perceive humans as predatory 'spirits', who make them fall ill and die. Though the relations are not limited to predation, they can be schematically represented (Fig. 2.1).

Viveiros de Castro concludes that whereas Western naturalism postulates a common nature, external to all representations, in which there are different cultures, which interpret the world differently (multicultural relativism), Amerindian perspectivism postulates a common culture, differentiated by particular natures (a multinaturalism), all of which are interpreted in the same way. This formulation and some of its derivations can be problematic.

Employing the notion of multiculturalism and contrasting Amerindian ontologies classifying them as multinaturalist is misleading—it projects pretensions to universalism characteristic of Christian Western ontologies onto Amerindian ontologies. Amerindian ontologies emphasize their geographical delimitations. They recognize the existence of other ethnicities, who do not share the pronoun *us*, because they belong to different environments, relate in different ways, and have different practices and moral codes—in sum, different cultures. This recognition implies that when encountering those others, the world becomes multicultural.

On the other hand, the idea that a perspectivist ontology implies a multinatural metaphysics 'in which there is no stable ontological ground'[26] does not follow. If the perspectives of the different beings are located in their bodies, then those bodies with their particular weapons and their affects towards other members of the community constitute a reality. Regardless of whether these bodies are perceived differently from different perspectives, the relative positions and their relations are held constant. Even if each being regards itself as human and interprets the

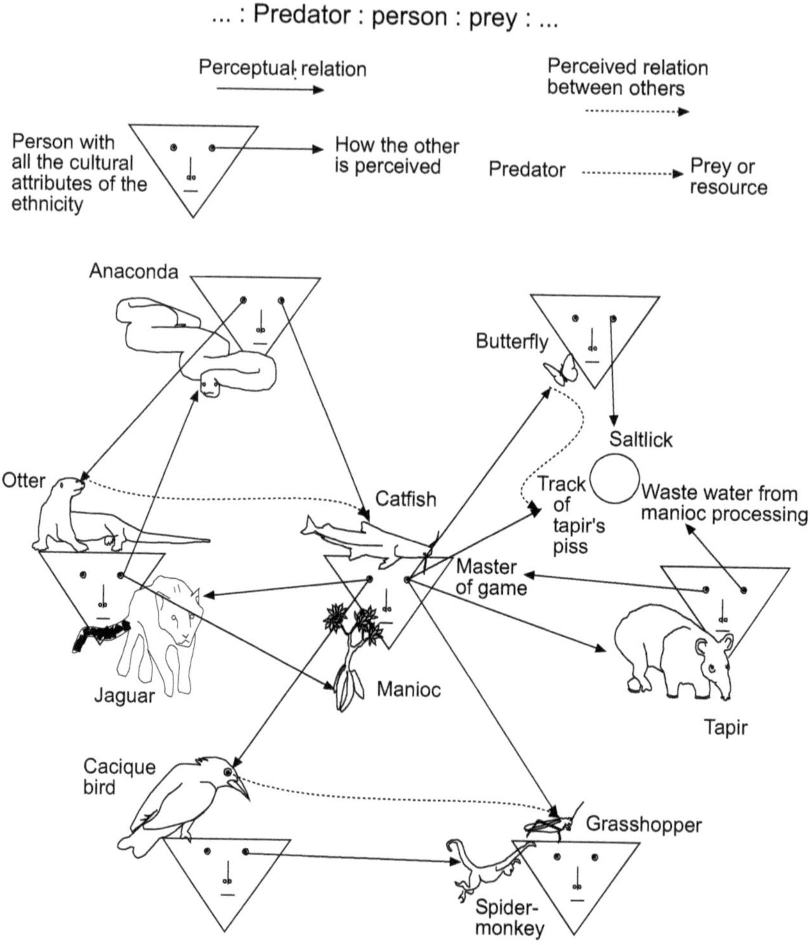

Fig. 2.1 Perceptual and ecological relations of Amerindian perspectivism

world as humans do, this is only possible because jaguars see as jaguars, fish as fish, and shamans employ the weapons of each of these beings to see like them. This is a stable ontological ground.

What is fascinating, what challenges Christian Western ontologies is that the ethnonyms designate all the members of an ecological community as bearers of that particular culture and as potential persons. This suggests that the pronoun denominates the *ecological community*. Making this adjustment, it can be suggested that in Amerindian perspectivist ontologies, what different species of beings share by virtue of belonging to an ecological community is not a metaphysical *internality*, as has been assumed by Descola (2013, 116), but rather a common field of meaning and interpretation, an environment of ecological relations, and the signs that make those relations possible.[27]

Willerslev (2004) points out that Viveiros de Castro's theoretic framework is too abstract and ought to be augmented with a theory of mimesis to address its practices. Like Amerindians, Siberian Yukaghir hunters are also perspectivists and employ insights from perspectival exchanges.[28] Elk hunters have to get rid of their human smell, camouflage their hunting intentions, and dress with special attire. Elk coats and skis covered in elk skin, which resemble the appearance and the sounds produced by elk moving through the snow, allow them to move like elk, and approach the way the elk perceives and is perceived. Other garments aim at mesmerizing elk with their beauty. The objective is to establish themselves as elk to the elk (or rather as human instead of as the predatory 'spirits' that are Yukaghir hunters to the elk people). With this elk shape and behaviours echoing those of the elk, the Yukaghir hunter sees people in elk and establishes an empathic bond, an emotional state, which allows him to honestly play a game of sexual seduction with the animal and if successful to kill it. However, sexual seduction is a risky game, in particular in the demanding rhythms imposed by pursuit. The hunter himself can become seduced by the prey, exhausted, and nauseated. He must avoid developing sympathy or else in proximity, his deceit could be discovered, causing the loss of the prey or even his own death. Worse, he could fall in love with the prey and become unable to hunt, suffer further debilitation, and even die or be killed. Or, equally dramatically, if he takes his role too seriously, he could metamorphose fully and lose himself in the world of the prey, forgetting he was ever a Yukaghir hunter.[29] The hunter should flirt through the doorway into the world of the prey, but refrain from crossing the threshold.

Stories about how hunters in close contact with their prey have metamorphosed and become lost in their world are frequent not only in the Siberian context, but also in the Amazon and the Andes. These stories are often recounted by those who, after a long sojourn in the prey community, were either told by their adoptive community or else realized that they were out of place.[30] These processes of involuntary metamorphosis have been connected to cultural hybridization. The latter also involves a process in which along and through close connivances with the other, garments, habits, and tastes are appropriated by individuals. This implies not only the incorporation of foreign aesthetics and ethics, but also deeper psychological and physiological transformations. Such transformations of individuals are often sought by the community to incorporate the other, in particular the dangerous other, be it the predator or non-native. It is a way by which people define themselves, not by isolation but through interrelation.[31]

This allows us to understand some of the offices of shamans. When a shaman dresses up, mimics the gestures, and adopts the postures of another kind of being, he becomes invested with some of its abilities and interprets the world through the habits of the mask, the body he has worn. He is then able to prolong his *I* into a common *we* and sees in the other beings an *us* instead of a *those*. Under those circumstances, the shaman is able to map in the other species the symmetry of habits, problems, values, and interests with which he is familiar. This is possible to a considerable degree because all members of an ecological community share in its overall complexity and its dynamics. On this common ground, the shaman converses and mediates with these other beings. Unlike the hunter, the shaman is an expert in crossing the threshold. The shaman has to behave as a native in the territory of the other to be able to enter into the other's home and share in their secrets, whether as an invited guest (friend) or as an uninvited victor (foe). Otherwise, there will be little record of the shamans' deeds back home.

2.4 Soulless Animism and Body-Full Spirits

Tylor defined animism as an early form of religion, a spiritualist philosophy of nature in line with idealism and in opposition to materialist philosophy. He assumed that there was a continuity in the ways the soul was conceptualized by 'primitives'—both ancients and non-Europeans—and 'civilized' Europeans. Primitive and modern Europeans might believe there must be unsubstantial, impalpable, and invisible soul or mind, which causes life and thought, animating an otherwise inert body, and assert that this soul independently possesses personal consciousness and volition. However, many Amerind 'animists' question the notion of the inanimate as unrealistic and highly problematic. They recognize that all beings and things are inherently undergoing multiple continuous processes of transformation, are embedded in complex interactions with their environment, and, in due course of their experience, may further acknowledge them as personae with a subjective stance, which requires negotiation. Contrary to long-held assumptions about the naïveté of animist beliefs, the more closely people interact with other beings/things and therefore the more knowledgeable they are about their behaviour, the more likely they are to acknowledge these beings as persons and to treat them accordingly.[32]

For animists, knowledge subjectifies instead of objectifying.[33] Knowing something is acknowledging its subjectivity—its particular way of being connected to a field of meaningful relations in the environment. This understanding of the world might be at odds with the idealized pursuits of modern science in its quest for universal deterministic laws, but it is in line with the complexities natural scientists face in reality, and with materialist understandings of complex phenomena. The antagonism between shamanism and materialist science might thus be misleading. Amerindian epistemologies mirror objectivist science, but they might, nevertheless, reach equivalent conclusions. While objectivist science attempts to fix deterministically what is known, only to discover further unknown frontiers, animists map the unknown frontiers, qualifying what is known. Animism leads us to consider that as biological beings, we are involved in meaningful interactions with other beings, interactions which shape us, and the environments we inhabit.

Amerindian perspectivist ontologies make a further corollary to animism; they recognize that culture is made up from the particular relations with different beings in an ecological community, and the signs mediating those relations, constituting a common and shared field of meaning and interpretation. They further recognize that each being is bound to see itself as the centre and measure of that field.

Our current understanding of animism suggests that the notion of the soul and, by extension, the notion of spirits are mistranslations. The question is: what was it that missionaries and explorers classified as souls and spirits in their first encounters with Amerind 'shamans'?

Notes

1. Flaherty 1989, 1992.
2. Eliade 1976.
3. Kehoe 2000.
4. Tylor 1870.
5. Tylor 1958 [1871], 9.
6. Tylor 1958 [1871], 13.
7. Descola 1996.
8. Durkheim 1969 [1915]; Lévy-Bruhl 1926, 1965 [1928].
9. Piaget 2009 [1927].
10. Guthrie 1995.
11. Tylor 1870; e.g. Durkheim 1969 [1915].
12. e.g. Green 1998, 1999.
13. Durkheim 1969 [1915]; Kehoe 2000.
14. Lévi-Strauss 1958, 208; Durkheim 1969 [1915]; Douglas 1984; Evans-Pritchard 1976; Kehoe 2000.
15. Prigogine and Stengers 1997; Prigogine 1998.
16. Latour 1987, 1999; Haraway 1985.
17. Schaeffer 2005.
18. Bennett 2010; Haraway 2003, 2008, 2016; Hird 2009; Tsing 2015.
19. Lévi-Strauss 1994 [1974].
20. Hallowell 2002 [1960], 23.
21. Fabiola Jara, personal communication 2014.
22. Hallowell 2002 [1960].
23. Guthrie 1995, 69.
24. Hallowell 2002 [1960].
25. Ingold 2000, 2011.
26. Viveiros de Castro 1998; Kohn 2015.
27. Kohn 2013, 163.
28. Willerslev 2004, 2007.
29. Willerslev 2004.

30. Fausto 2007; Santos-Granero 2009; Willerslev 2004.
31. Santos-Granero 2009; Viveiros de Castro 2011.
32. Willerslev 2007, 16; Bird-David 1999.
33. Viveiros de Castro 2004.

Bibliography

Århem, Kaj. 1993. Ecosofia makuna. In *La selva humanizada: ecología alternativa en el trópico húmedo colombiano*, ed. François Correa, 109–126. Bogotá: ICAN/FEN/CEREC.

Bennett, Jane. 2010. *Vibrant matter: A political ecology of things*. Durham: Duke University Press.

Bird-David, Nurit. 1999. "Animism" revisited: Personhood, environment, and relational epistemology [and comments and reply]. *Current Anthropology* 40: 67–91.

Descola, Philippe. 1996. Constructing natures: Symbolic ecology and social practice. In *Nature and society: Anthropological perspectives*, ed. Philippe Descola and Gisli Palsson, 82–101. London: Routledge.

———. 2013. *Beyond nature and culture*. Trans. J. Lloyd. Chicago: University of Chicago Press.

Douglas, Mary. 1984. *Purity and danger: An analysis of the concepts of pollution and taboo*. London: Routledge.

Durkheim, E. 1969 [1915]. *The elementary forms of the religious life*. Trans. Joseph Ward Swain. New York: Free Press.

Eliade, Mircea. 1976. *El chamanismo y las técnicas arcaicas del éxtasis*. Trans. Ernestina de Champourcin. Bogotá: Fondo de cultura económica.

Evans-Pritchard, E.E. 1976. *Witchcraft, oracles and magic among the Azande*. Oxford: Clarendon Press.

Fausto, Carlos. 2007. Feasting on people: Eating animals and humans in Amazonia. *Current Anthropology* 48 (4): 497–530. https://doi.org/10.1086/518298.

Flaherty, Gloria. 1989. Goethe and shamanism. *MLN* 104 (3): 580–596. https://doi.org/10.2307/2905047.

———. 1992. *Shamanism and the eighteenth century*. Princeton: Princeton University Press.

Green, Edward C. 1998. Etiology in human and animal ethnomedicine. *Agriculture and Human Values* 15 (2): 127–131.

———. 1999. *Indigenous theories of contagious disease*. Oxford: Altamira Press.
Guthrie, Stewart. 1995. *Faces in the clouds: A new theory of religion*. New York: Oxford University Press.
Hallowell, A.I. 2002 [1960]. Ojibwa ontology, behavior, and world view. In *Readings in Indigenous Religions*, 18–49. London: Continuum.
Haraway, Donna Jeanne. 1985. *A manifesto for cyborgs: Science, technology, and socialist feminism in the 1980s*. San Francisco: Center for Social Research and Education.
———. 2003. *The companion species manifesto: Dogs, people, and significant otherness*. Chicago: Prickly Paradigm Press.
———. 2008. *When species meet*. Minneapolis: University Of Minnesota Press.
———. 2016. *Staying with the trouble: Making Kin in the Chthulucene*. Durham: Duke University Press.
Hird, Myra J. 2009. *The origins of sociable life evolution after science studies*. New York: Palgrave Macmillan.
Ingold, Tim. 2000. *The perception of the environment: Essays on livelihood, dwelling and skill*. New York: Routledge.
———. 2006. Rethinking the animate, re-animating thought. *Ethnos* 71 (1): 9–20.
———. 2011. *Being alive: Essays on movement, knowledge and description*. Oxford: Routledge.
Kehoe, A.B. 2000. *Shamans and religion: An anthropological exploration in critical thinking*. Long Grove: Waveland.
Kohn, Eduardo. 2013. *How forests think: Toward an anthropology beyond the human*. Berkeley: University of California Press.
———. 2015. Anthropology of ontologies. *Annual Review of Anthropology* 44: 311–327.
Latour, Bruno. 1987. *Science in action: How to follow scientists and engineers through society*. Cambridge, MA: Harvard University Press.
———. 1999. *Pandora's hope: Essays on the reality of science studies*. Cambridge, MA: Harvard University Press.
Lévi-Strauss, Claude. 1958. *Structural anthropology*. Paris: Plon.
———. 1994 [1974]. *Antropología Estructural*, vol. I. Barcelona: Altaya.
Lévy-Bruhl, Lucien. 1926. *How natives think*.
———. 1965 [1928]. *The soul of the primitive*. Trans. Lilian A. Clare. London: Allen & Unwin.
Piaget, Jean. 2009 [1927]. Children's understanding of causality. *British Journal of Psychology* 100 (S1): 207–224.

Prigogine, Ilya. 1998. *Exploring complexity*. New York: Freeman and Company.
Prigogine, Ilya, and Isabelle Stengers. 1997. *The end of certainity: Time, chaos and the laws of nature*. Nueva York: The Free Press.
Santos-Granero, F. 2009. Hybrid bodyscapes: A visual history of Yanesha patterns of cultural change. *Current Anthropology* 50 (4): 477–512.
Schaeffer, Jean-Marie. 2005. La thèse de l'exception humaine. *Communications*: 189–209.
Tsing, Anna Lowenhaupt. 2015. *The mushroom at the end of the world: On the possibility of life in capitalist ruins*. Princeton: Princeton University Press.
Tylor, Edward B. 1870. The philosophy of religion among the lower races of mankind. *The Journal of the Ethnological Society of London (1869–1870)* 2 (4): 369–381.
———. 1958 [1871]. *Religion in primitive culture*. New York: Harper & Brothers. Original edition, as Chapters XI–XIX of *Primitive Culture*.
Viveiros de Castro, Eduardo. 1998. Cosmological deixis and Amerindian perspectivism. *Journal of the Royal Anthropological Institute* 4 (3): 469–488.
———. 2004. Exchanging perspectives: The transformation of objects into subjects in Amerindian ontologies. *Common Knowledge* 10 (3): 463–484.
———. 2011. *The inconstancy of the Indian soul: The encounter of catholics and cannibals in 16th-century Brazil*. Trans. Gregory Duff Morton. Chicago: Prickly Paradigm Press.
Willerslev, Rane. 2004. Not animal, not *not*-animal: Hunting, imitation and empathetic knowledge among the Siberian Yukaghirs. *The Journal of the Royal Anthropological Institute* 10 (3): 629–652.
———. 2007. *Soul hunters: Hunting, animism, and personhood among the Siberian Yukaghirs*. Berkeley: University of California Press.

3

First Contacts with Amerindian Shamans and Their 'Spirits'

This chapter examines two early encounters with Amerinds in the Caribbean: the Taino and their closely related enemies the Callinago, with their 'shamans' respectively the *buhities* and the *boyáicou*, and the entities with which they dealt. It explores the descriptions of the *zemes*, entities which missionaries translated as gods, demons, or evil spirits, and other beings with which these physician-magicians interacted. It analyzes these ontologies in their own terms, gathering evidence to argue that these entities were acknowledged with characteristics of living and social organisms, which were part of an ecological community, including humans. It analyzes the accounts of the practices and the claims of these shamans, showing how they involved biological and emotional effective means of managing health and the environment. Finally, it discusses the politics of the encounter with missionaries and explorers, the interests, the process of appropriation, translation, and purification of their ontologies.

C. E. Giraldo Herrera, *Microbes and Other Shamanic Beings*,
https://doi.org/10.1007/978-3-319-71318-2_3

3.1 The Hispano-American Encounter, from the Middle Ages to the Renaissance

The earliest descriptions of practices resembling shamanism in the Americas were drawn by Friar Ramón Pané and by Captain Gonzalo Fernández de Oviedo. Pané was a Hieronymite friar commissioned by Columbus to learn about the antiquities, beliefs, and idolatries of the Tainos inhabitants of the Hispaniola island, between the second and the third Columbian Voyages (1494–1498).[1] He lived amongst various Taino communities, and while evangelizing, he recorded their myths, producing the earliest observations about Amerindian culture. Oviedo, on the other hand, was a naturalist, historian, and a chronicler of Indies. He participated in various expeditions against the peoples of *Castilla de Oro* (the coasts of today's Colombia, Panama, and Nicaragua) and became mayor of the fortress of Santo Domingo. Early on, he professed doubts regarding the humanity of Amerind peoples. Nevertheless, he drew a rich intelligence on their culture, often demonstrating admiration towards his adversaries. His views tend to become more amiable towards Amerinds as his *General and Natural History of the Indies* progresses.

The friar draws a hesitant portrait of the beliefs of the Tainos. After a brief introduction, *Pané* mentions the zemes (*çemies*),[2] identifying them as idols of the gods of the Taino:

> They believe their god is in the sky, is immortal and nobody can see him, has a mother but no beginning, and they call him Yúcahu Bagua Maórocoti and his mother they call Atabey, Yermao, Guacar, Apito and Zuimaco, which are five names.[3]

His initial depiction of the most important Taino deity suggests some features in common with the Christian God. Like the latter, Yúcahu Bagua Maórocoti lives in the sky and is immortal and invisible, and although it has a mother, it has no beginning. The friar seems to be seeking cultural analogies that facilitate the process of conversion. However, after providing an account of Taino mythology, with which we will deal more deeply in the **Third Part** of the book, he swiftly disqualifies it as a set of naïve beliefs, excusable on account of the ignorance of the Taino regarding the 'true faith', thereby justifying the enterprise of their conversion. Later on, Pané will further disqualify zemes as 'idols or more properly as demons made by the naïve who believe on them'.

Pané remarks that every Taino had many zemes, some made of wood and many others made of stone, but all different. He does not specify features distinguishing an object as *zeme*; seemingly the term could encompass a variety of objects ranging from stones of various sizes, vessels, woven cotton artefacts, and wooden or stone figures with human and/or animal shapes.

Turnip-shaped stone zemes talked and gave advice in visions, after people fasted, consuming and taking baths with herbs, like *güeyo* and *cohoba* powder. Other zemes were the source of what the Taino ate: fist-size three-pointer zemes made manioc grow, other zemes regulated the abundancy of fish and game.[4] Arróm suggests that three-pointers represented Yúcahu Bagua Maorócoti, whose name has been translated as lord-of-manioc and the sea, without grandfather. Various authors argue that these figures were interred in manioc fields and urinated upon to fertilize the ground.[5] There are a number of archaeological artefacts which correspond to Pané's description of a three-pointer.[6] Three-pointers are frequently made out of marine-related materials such as limestone, seashell, coral, and mammal bone. One of the most highly distributed zeme materials is calcirudite, a variety of limestone produced by microbial activity, which is only found in a few locations in the Caribbean.[7] According to Pané still other zemes were employed to contain the bones of the ancestors.

Zemes could have names, Boynayel and Mároju, a couple of anthropomorphic arm-length stone zemes portrayed with their hands bound were kept in a cave and sweated (presumably condensing water). Respectively they propitiated calm and windy weather. Likewise, Guabancex, a female zeme made of various stones, when angered produced hurricanes and floods. She controlled the wind and the rain with the assistance of two other zemes: Guataúba was her herald and commanded on her behalf multiple other zemes to collect winds. Coatrisquie was her collector, and he gathered waters from the mountains to destroy the country. Zemes were whimsical. One made of wood and called Corocoto frequently fled and had intercourse with the women of the village. Various children of the village who had been born with 'two crowns'[8] were said to be sons of Corocoto.

Pané[9] also describes the process whereby a zeme was found and made. If a Taino saw a tree moving more violently with the wind than its neighbours, he would call for a *buhitie*.[10] Pané describes *buhities* as medics, who gained authority from healing their patients and employed the inebriating power of inhaled *cohoba*[11] powder to fool people into believing whatever they wanted. Following the call, the *buhitie* would hold a *cohoba*, a procedure in which the participants would become inebriated after inhaling the substance. The *buhitie* would inquire who it was and what it wanted to be done, whether it wanted a zeme to be carved from the tree, to get a shrine on its behalf, or whether it wanted a state of land to be reserved. After the *cohoba* concluded, the desires the zeme expressed to the *buhitie* would be obeyed, and if required, the tree would be felled and a suitable figure would be carved out of it, a shrine would be built for it, or a state of land would be reserved on its behalf. According to the friar, the Tainos normally complied with these requests, partly for fear of the consequences. The zemes would cause disease if they were not properly kept, or if their requests or advice was not followed. On the other hand, the Tainos also complied with the requests of the zemes because chieftains competed with one another over the power of their zemes, the sumptuousness of their shrines, and the size of their zeme's states.

Pané[12] describes the healing procedures of the *buhities*. In preparation for the *cohoba*, the medics would darken their faces with soot. When they arrived at the communal house where the patient resided, it would be emptied, except for the latter and one or two principals. The *buhities* had to follow the same fasting regimes as their patients. They had to purge themselves employing *cohoba* powder, *güeyo* or other leaves and had to imitate the patient's expressions. The *buhitie* would be the first to ingest *güeyo* and other mashed leaves, and to vomit. Then he would start the *cohoba*. After inebriating himself and the participants, he would sing the *areytos* or mythical songs and lighten a torch. The *buhitie* would observe and palpate the patient. He would talk to the zeme causing the disease, requesting that it went away, and blow on his hands as if to throw it to the wind. After that, the *buhitie* would start sucking the afflicted part of the patient. He would start coughing and expressing disgust. Finally, the *buhitie* would produce out of his mouth objects such as a bone, a piece of meat, or a stone which, he would claim, was the material cause of their illness and had either been ingested by the patient or been planted by their zeme in retaliation for some failure or disobedience.

However, some of the objects extracted, such as stones, were taken to be zemes and were highly treasured. These stone zemes assisted women in giving birth. These zemes would be wrapped in cotton, placed in a basket, and, like other zemes, would be fed with the same products the owner ate. This was the source of hilarity for the friar who considered that being made of inert matter zemes could not possibly consume any such nourishment, and thereby the ritual was futile but also innocuous and inexpensive.[13]

Albeit the *buhities* had considerable room to play schemes on their patients and on the wider community, if their patients died, they could be held responsible. Though the friar claims that few people dared confront a *buhitie*, if the dead was a principal or had many relatives, the family would hold a *cohoba* addressing enquiries towards the corpse, which in that circumstance was treated like a zeme. Moreover, if the *buhitie* had failed to fast he would suffer rashes, further attesting to his lack of commitment. If the family concluded that the *buhitie* had not kept the required diet, his or her lynching would ensue. Nevertheless, even in those circumstances, *buhities* turned to be extremely difficult to kill—supposedly their *zemes* came to their assistance in the form of snakes and revived them. This was used as an excuse to justify various sorts of torture and desecration of the body of the *buhiti*.[14]

Besides describing zemes and *buhities*, Pané mentions the Taino differentiated between the 'spirit' of the living, the *goeiza*, and that of the death, the *opia*. He does not elaborate on the *goeiza*, but indicates the *opia* fed on guava fruit, and seeking sexual encounters, they would appear in 'human form' to people walking at night alone through the forest. *Opia* could be recognized by their lacking a navel and would disappear in the middle of a loving embrace.[15]

Oviedo[16] seconds Pané, describing zemes as images of the devil. However, he warrants that these beliefs were very deeply rooted amongst the 'naturals', declaring it unlikely that they could be uprooted. He remarks that the naturals, both in the island and in the continent, would carve, sculpt, model, or paint these images everywhere, in the stools (*duhos*) on which their principals sat, in the metal breast and nose-plates with which they adorned themselves; they would even ink them on their own skin. He describes zemes with many heads or tails, with ferocious yaws and humongous fangs, or with oversized ears and the fiery eyes of dragons or snakes, but always varied and different.

In Oviedo[17] we also find a description which matches the consultation of the corpse as a zeme, but performed on a Spaniard. The Spaniards under the leadership of Ponce de Leon had settled in St. Johan or Borinquen (Puerto Rico), spreading over the island and imposing tribute on its inhabitants. The natives although unwilling to submit were fearful of their military forces. Oviedo claims the fear was partly inspired by beliefs in the immortality of Spaniards. However, he has previously described various insurrections in which the Indians had murdered many Spaniards, only to be later overwhelmed by their forces. Nonetheless, to verify the supposed immortality, a cacique called Urayoan lured a young Spaniard named Salçedo, had him ambushed and drowned in a river. After the man died, the natives placed the corpse on the banks of the river and started questioning him. For three days the 'experiment', as Oviedo calls it, continued allowing the corpse to rot in the water. Still unsatisfied, Urayoan had the enquiries proceed for a few days longer. Having confirmed the 'mortality of Spaniards', the natives rebelled. Nonetheless, after considerable bloodshed, the Spaniards controlled the revolt.

Oviedo[18] regarded *buhities* mainly as diviners or magicians who adored these zeme 'devils'. He noted that *buhities* always carried zeme images and had them painted on themselves. Moreover, they themselves were identified as zemes. Nonetheless, he recognized that they were great herbalists who knew the properties of many trees, plants, and herbs. Because they healed many through these arts, they were greatly revered and respected. For this reason, they were taken for saints, fulfilling the same roles as priests amongst Christians. Oviedo did not stop in that comparison but refers to Pliny through Isidore of Seville, and likens *buhities* to the magicians of Persia. He pointed out that the arts of medicine, religion, and mathematics (i.e. also astrology and divination) were as tightly linked together for Amerinds, as they had been for the ancients of the East.

Buhities interpreted the advice of zemes regarding illness, war, or whatever a person desired.

However, to Oviedo's bafflement, the *buhities* did not lose credibility even when their forecasts did not come about. Because, they argued, zemes could change their minds or could be pursuing their own whims or self-interests. This, to his surprise, did not affect but rather redouble the verisimilitude of the zemes with which they interacted.

Pané and Oviedo had radically different formations, positions, and interests. While the latter sought reasons to justify a military campaign with its exploits, the former saw the potential for conversion in the naïveté of the Taino. The friar is one of the earliest Europeans in the Americas, and in many ways he is a representative of medieval European culture. He considers zemes as inert matter and the procedures of *buhities* as innocuous scams, subsuming medicine to the healing power of suggestion. Thus, he reduces the ontology of *buhities* to the realm of religion. However, his work presented the Church with means to acquire power and lands by co-opting the position of the *buhities* in Taino society and replacing zemes with the images of saints.

Oviedo, on the other hand, represents a second generation, a response to the Encounter. Oviedo is a man of the Renaissance, a personal friend of Leonardo da Vinci. He is the fruit of a culture which is already recovering the work of classical natural historians like Pliny, albeit only through secondary sources. Oviedo is able to acknowledge the naturalistic knowledge of *buhities*. Nonetheless, he presumes there must be a natural order to the world (of which only humans are exempt). Thus, he does not seem to understand that a non-human entity behaves in unpredictable ways, much less how such capricious behaviour demonstrates the intrinsic power of zemes. Both the friar and the natural historian however believe in the existence of the devil and recognize in zemes its signature.

From Pané and Oviedo, we gather that the Taino recognized a highly varied multiplicity of *zemes*. These entities were regarded as powerful agents with self-interests, limited but adaptable knowledge and whims, that is with their own subjective intentionality, and also with social hierarchies. These entities were perceivable through visions and were biologically affected and effective. Although they were not limited by bodily boundaries, they could be partially embodied in stones, artefacts, plants, animals, humans, or their corpses. Their embodiment was recognized by extraordinary differentiating behaviours. They influenced health and local ecologies. They had particular nourishment requirements, and their capacities were affected by fasting, by physiologically active substances, as well as by meaningful, emotionally loaded, performances developed by *buhities*.

If we take into account both Pané's version of the consultation of corpses as zemes, and the often quoted[19] episode narrated by Oviedo about the drowning of a Spaniard to experiment on Spaniard 'immortality', we may conclude that these consultations involved physical empirical methods and were guided by materialistic rather than supernatural concerns. The expression of the zemes was correlated with the process of rotting undergone by those corpses. From a biological perspective, the Urayoan's experiment on Salçedo's corpse revealed whether Spaniards were hosting or were susceptible to particular microorganisms.

Pané recognizes *buhitites* as successful medics, but shows only the means of their scams. Oviedo portrays them as diviner magicians, but acknowledges the prowess of their medical and botanical knowledge. Their combined testimony indicates that *buhities* were accomplished medics and influential environmental managers. They were considered as zemes, and facilitated the communication with other zemes through the communal consumption of substances like *cohoba* and *güeyo*. Through their interpretation of the will of zemes, *buhities* were in the capacity of establishing reservoirs and regulating cultivation, fishing and hunting, and managing ecological relations. Advising on personal and communal matters, such as war, they were also highly politically influential.

3.2 The French Encounter with Shamanism that Preceeded the Enlightenment

Roughly a century after Pané and Oviedo, Raymond Breton and three other French Dominican friars were commissioned by Richelieu to facilitate relations and forge alliances with the Antillean Caribs, gather information on their language and culture, and convert them to Catholicism.[20]

Breton arrived at the Antilles in 1635 and lived amongst the Callinago of Guadalupe and Dominica for over 20 years. With no reference material at his disposition, he compiled a Carib-French dictionary, with extensive comments of reference to their culture, a French-Carib dictionary, and a Carib catechism.[21] The dictionaries were destined to the formation of missionaries in the Caribbean; they were the supporting framework for the Carib catechism. Historians of the region, like du Tertre[22] and de Rochefort,[23] although themselves spending time as missionaries in the isles and publishing earlier, acknowledged the importance of Breton's testimony for fleshing up their accounts of Carib culture. A hundred years later, Labat would draw information from du Tertre. However, because they were destined for a more general public and perhaps were under closer inquisitorial scrutiny, they often upheld more dogmatic positions. Nevertheless, their personal experiences, access to alternative sources, and the narrative structure of their work often render supplementary insights and allow us to regard in perspective the views underlying the development of scientific knowledge from a 'natural history' which was already paired with a 'moral history'.

According to Breton, the Insular Caribs, or Galibis of the Antilles, referred to themselves by the name Callinago, meaning good people or people of peace. They claimed to be descendants from the Continental Gallibis (from Surinam and Guyana) who had invaded the islands, wiped out *Arouage* men, and retained *Arouage* women as their own wives. Both men and women preserved the use of their respective dialects.[24] Nevertheless, the Callinago registered in the dictionary only has a few Carib elements; by large it belongs to the Arawakan family and is closely related to Taino.[25] The peoples designated by Europeans as *Arouage*, the sworn enemies of the Callinago, were called *Etoútou*. This word did not necessarily have a specific ethnic affiliation, but probably meant non-Callinago person (from *Amoútteti*: different).[26] Christians were likewise denominated *etoútou noubi*, probably meaning latest other.[27] Because of the linguistic filiations, and Breton's claims not to have used external sources, we can assume that the resemblances between his observations and those of Oviedo and Pané represent commonalities with the Taino and continuities within Callinago culture.

3.2.1 *Acansáncou* or *Oupoyem*: Spiritual Entities

The explicit purpose of Breton's dictionary is to develop a context, a framework for the translation of the Catechism. To do this, he develops a spiritual vocabulary. However, the Callinago 'knew only what they could see and refused to acknowledge what they did not see or did not use'.[28] Breton indicates that abstract notions are scarce in Callinago, as well as words for spiritual and political matters. Despite Callinago bias towards empiricism and the 'insufficiencies' of their language to express spiritual matters, Breton reduces a plethora of Callinago terms into the limited religious vocabulary of spirits, souls, devils, god, magic, sorcery, spells, and curses. Du Tertre and Labat reduced matters further, subsuming a wide variety of Callinago terms under the rubric of mápoya or devil.

Breton introduces the notion of spirits[29] to translate *acansáncou* or *acàmbouée* (m.) and *oupoyem* (f.). *Acansáncou* or *acàmbouée*[30] can be associated with *acàmboni*: understanding or science and with *acáli*: form, figure, or soul. He translates the quality of being spirited, of 'having spirit', as *lioüánni* (force or strength), which was related with *ioüánni*: heart, life, worth, and soul. In turn, *Ioüánni* could be related with *Iaoüa* shadow and *Iaoüátêna* image.[31] Breton also translates soul as *ánichi* (heart), a noun which together with particular verbs denoted envy, will, or desire, but could refer specifically to the pulse, as in *noúbaco ánichi*, or soul of the hand. Besides the plurality of souls, these notions introduced other problems:

> *mébechoüatïtiu tébou oüekeliem*: men 'don't turn into stones.' Our naturals have not heard about Pythagoras. Nevertheless, they seem penetrated by his metempsychosis.[32]

Metempsychosis is the belief in the transmigration of the souls across different animal species. Seemingly, for Callinago the transmigration was not limited to animals but included the mineral kingdom. Moreover, processes such as the fermentation of wine or *tapouchicani* were also understood by the Callinago as the result of the agency of such spirits. Literally, this word was translated as 'spirit turmoil or the spirits of wine jumping'.[33]

Like *acansáncou*, *oupoyem* was a general category for spirit. Oupoyem was related with matter (*oubó*, *tóboyem*, or *toubóyana*) and with *boyé*, the word for shaman.[34] According to Breton, this category also included evil spirits such as the *mápoya* (pl. *mápoyanum*), *oumécou*, and *couloúbi*. The friar claims the Callinago lacked any history about their origin or that of the world, but had numerous tales about these entities. Nonetheless, these spirits were feared and abhorred; they were the cause of many diseases, mishaps, and deaths. Breton claims to have taken advantage of these attitudes towards *mápoyanum* to translate the notion of devil.[35] He contrasted the notion of *Mápoya* with the Callinago notions of *Çemijn*[36] *f.* (pl. *çeméignum*)[37] and *Ichéiri m.* (pl. *ichéiricou*), which he employed to translate God.

The *çeméignum* or *ichéiricou* were the 'pretended gods' of the naturals. They were *aráliracaútium*, that is, they descended from *Rálou Ema*, the Great Path above in the sky.[38] Amongst *ichéiricou* there were children, youngsters, adults, and elderly and males and females. Each *ichéiricou* had endless sons and grandsons, which they took with themselves to eat and drink the offerings made in their honour. They also belonged to nations corresponding to those of humans, such as the rival *Arouage*.[39] The Callinago honoured *ichéiricou* because they granted them good health and success in war or made the manioc grow. They also feared the *ichéiricou* because they caused hurricanes and intervened in diseases like those of the joints or gout. Moreover, the Callinago refrained from killing *loumakachítina*, wild animals that behaved like domestic ones, recklessly approaching human dwellings. These animals were thought to belong to the *ichéiricou*, who would retaliate by killing them.[40]

Rochefort points out that Caribs normally would not claim that the *Ichéiricou* were creators of the world. But they would retort to the claim that God created Heaven and Earth:

> True, your God made the Heaven and Earth of France and causes the wheat to grow there, but our gods made our Country and cause our manioc to grow.[41]

He also points out that for the 'savages' the soul was a subtle substance, rather than something immaterial or invisible. They thought each person had as many souls [*ánichi*] as places where they could feel their pulse. Of all these souls, the principal was in the heart [*ioüánni*]. Upon death, this one was carried to the sky by its *ichéiri*, to live a human-like life as an *ichéiri* amongst the *ichéiricou*. The spirit of the deceased could also take abode in its bones, which they wrapped in cotton and treated as oracles. They would bewitch their enemies wrapping these bones together with something that belonged to the latter, to cause disease or mishap, or death.[42]

Other souls in the body, Rochefort adds, would turn into *oumecou*: evil spirits that go to reside by the seaside and cause mishaps for vessels at sea, or go to live in the forests as *maboya*.[43] Foul smells in the forest identified the presence of a *maboya* at a particular site, and the naturals rushed to leave those places upon noting their presence. They also attributed the name *maboya* to certain plants and mushrooms of ill scent, and in their first encounters with Europeans, the naturals also gave this name to guns, and assumed gun-powder to be seeds, which they proceeded to plant.[44]

A century later, Labat claims that the Caribs do not really have notions for God or for spirits but solely abstract notions for good and evil. Nonetheless they are constantly tormented by the *maboya* or the devil. For that reason, they sought to have a Christian living in their communal house. If that was not possible, then a cross should be hung inside the communal house. Nevertheless, they were superstitious, and should it happen that there was bad hunting or fishing, they would claim that the cross had angered the devil, and would throw it down and tear it apart.[45]

Although notions such as *ichéiricou*, *mápoya*, and *ioüani* provided missionaries with the means to translate fundamental religious concepts, adopting these notions also implied disputing authority over them and coming into confrontation with the specialists and the procedures dealing with these entities. Du Tertre claims that all the ceremonies of the Caribs in particular those of the *boyé,* their priests or magicians summoned the *mapoya* or the devil.[46] De Rochefort and Breton deny this accusation, remarking that for the Callinago, the difference between *icheiricou and mápoya* was as profound as for Christians the difference between God and the devil. Breton adds the Callinago considered unbelievable that anyone would make offerings to a *mápoya* or demon. Nevertheless, they both consider the *ichéiricou* as spirits of darkness, equivalent in every sense to *mápoya*, and worshipers of *ichéricou* as equal to worshipers of *mápoya* or devils.[47]

3.2.2 *Boyáicou*: The Rival Magicians and Their Idolatries

De Rochefort claims that each Carib had his or her particular *ichéiricou,* to which they made offerings of cassava breath and *oiucou* (a fermented beverage made from manioc), when they felt these gods had cured them from some disease or preserved them from some mishap. However, to summon the *ichéiricou,* in order to seek advice in war, vengeance for some mishap or disease caused by another person's sorcery, to be cured, or to drive away *maboya,* they required the offices of a *boyé,* a magician.[48]

Breton, on the other hand, claims that the *ichéiricou* were not gods of all the naturals but solely of each particular *boyé.* He describes the *boyé* (pl. *boyáicou*) as physicians, priests, or magicians who, with the assistance of their *ichéiricou*, cured and gave advice.[49] The *boyé* presented their *ichéiricou* with every novelty arising in the community. Whenever the Callinago went to war or had sick people, they prepared a *matoútou,* an offering, and called for the *boyé* to present it to his or her *ichéiri.*

The *boyé* arrived to the communal house, sometimes with an assistant, and intoned a *lerèmerícani,* a song to make the *ichéiricou* descend. Breton refused to translate or even to write these songs, for fear of reproducing the words of the devil. While singing, either the *boyé* or the assistant would blow tobacco smoke. The *boyé* presented the *ichéiricou* with a hammock to sit in and a table full of offerings to eat and drink. The friar assumed this was a mock-up of the Christian ritual. However, if we attend to Rochefort's notes on diplomacy and courtesy, it becomes evident that these ceremonies replicated the courtesies reserved for the reception of most notable guests.[50]

The *ichéiricou* commanded the fire to be put down and all windows to be sealed, and then descended. The friar interpreted this order as evidence that the *ichéiricou* were spirits of darkness. Alternatively, a *boyé* claimed: '*hamánhatina*': I am flying. He would further ascertain that his gods elevated him before everyone's eyes, and he flew to the sky of the moon. The other participants confirmed his fit, exclaiming with admiration '*kárênati*' he has wings.[51] The friar assumed this was the effect of an enchantment and was not the least compelled by the *boyé's* description of heaven as 'half-open stones distilling water everywhere', which did not meet his expectations for a mystical journey.

Either having the *ichéiri* descend to the *carbet* or being lifted to the sky, the *boyáicou* entered into conference with *ichéiricou,* who gave them advice in all sorts of matters, from personal grievances to war. They also indicated to the *boyáicou* which specific herbs to take for each particular disease and would assist them sucking (*kachoulocátium)* the arrowheads, stones, or spines which caused disease. Breton dismissed these procedures as ludicrous scams, for which the *boyáicou* charged outrageous fees.[52]

Moreover, the friar continues critically: although the *ichéiricou* supposedly drank all the wine, moving the pots and making them sound as if they were empty, the following day the pots would be miraculously full again, causing general amazement. Elderly people and principals who had refrained from sexual intercourse during the previous night would become drunk. Again, Breton assumed this was a scam. Moreover, the *boyáicou* were imitating the Christian liturgy.

To make matters worse, in the eyes of the Breton, children would be consecrated to some particular *ichéiricou* so that they too became *boyáicou.* Aspirants would train under an established *boyé,* carrying out fastings, after which the *boyé* made his or her *ichéiricou* descend.[53] These *ichéiricou* gave aspirants their own gods or goddesses, with an appearance and voice corresponding to those of the aspirant.

Breton developed a tense and highly antagonistic relation with the *boyáicou.* Not only he considered them as quacks and worshipers of the devil, he also regarded them as rivals: sought to prevent promising and influential converts from undergoing *boyé*-apprenticeship and ordered prayers attempting to frustrate their reception of the *ichéiricou.*[54] Moreover, wearing the stole and holy sacraments in hand, he surreptitiously introduced himself into their procedures, attempting to disrupt them, for instance, by bringing light into spaces that had been darkened.

On one of these incursions, Breton found the *boyé,* a woman, wailing and screaming, throwing herself to the floor and crying. The spectacle continued for over a quarter of an hour, and all the participants were so enthralled and absorbed that the friar decided to abandon the place. As he was about to leave, the *boyé* caught sight of him, and the 'evil spirit' mounted in rage:

> Hurry, hurry, have him behold and tied down for fear that he may escape. May I eat his head, shoulder, feet, excrements and all, crush him, smash him to pulp, and devour him whole.[55]

The threat of *boyé* was not to be taken lightly. In various entries, Breton mentions the feasts made by the Callinago on *Arouage* prisoners. De Rochefort adds further gruesome detail: war prisoners were cut while they were alive, and captives would demonstrate their bravery defying their enemies through the process. The captives boasted of their own acts of cannibalism against the captors' ancestors, revealing how they were going to consume meat and grease from their own kin. After enduring torture, the captive would be killed by a single blow of a mace. The meat would be seasoned and cooked. The grease from the corpse would be saved and treasured as a 'ceremonial seasoning'.[56] Breton never dared landing on St. Cristoph, because two missionaries had already been killed there. Moreover, two of his own Guadalupean converts were killed after returning from Paris. One of them was murdered despite having authorization from the community for travelling. However, Breton survived the incident mostly unscathed. Long afterwards the Callinago teased the friar, imitating the *boyé's* wrathful anthropophagous threat, with its sexual undertones,[57] while laughing.

The *boyé* left durable marks on the friar's memory. In his entry for the Callinago, he would vindicate them from the accusations of cruel anthropophagy: 'I have no reason to complain of their cruelty, rather perhaps, I would complain of their sweetness towards me'. And concluded paraphrasing the episode: '(Oh! How sweet cruelty, to be crushed, devoured and torn apart for defending a God!)'.[58]

3.2.3 Naturalizing Callinago Medicine, Trivializing Sorcery, and Superstition

Although the Callinago feasted on Arouage and European prisoners, they felt a profound repulsion towards eating animals which they themselves had fed and would not even taste the eggs produced by hens they had raised.[59] They also followed numerous food prohibitions: they refrained from consuming the meat and the grease of swine, parrot, manatee, eels, as well as other birds and fish to prevent their progeny from inheriting characteristics from these animals. Different communities had different prohibitions. While Breton details Callinago techniques for hunting and cooking turtle, de Rochefort's Caribs refrained from the meat of turtle to avoid acquiring its stupidity. Moreover, they would neither eat salt nor preserve their meat with it: 'when they saw Europeans salting their food, they would claim: you hasten your own death'—[60]a statement of 'superstitious ignorance' for which de Rochefort ridicules them. Breton finds their reasons equally ludicrous but justifies their praxis noting that 'the naturals who violated them, eating as Europeans did, were prone to illness and short-lived'.[61]

The Callinago would undergo severe fasting in various special circumstances, such as the birth of a first-born son, the capture of an enemy, or the death of a relative. Fasting normally involved five days without eating or drinking and four days drinking only boiled manioc, after which they were allowed to eat cassava bread for another three months. Following this period of time, they would still abstain from certain meats.[62] The circumstances that required fasting were such novelties as to be presented to the *Ichéiricou*. The birth of a first-born son and the capture of an enemy are perilous circumstances, which presuppose being preserved from mishap and thus the favour of the *Ichéiricou*. On the other hand, disease and death were generally attributed to the *Ichéiricou* of enemy *boyáicou*, to *mápoyanum*, or to sorcerers. Deciding which of these agents was responsible would determine whether a war or witch-hunt was called for. These were clearly functions which were required of the *boyáicou*. Breton does not specify the meaning of the various words that designated fasting: *nenemaiíntina, huénemabouicyéntina, nihuenemaíntina enemapátina, iuénemepátina.* Some of them seem related with words Breton employs to translate dream (*ihuénnête)* and fantasy (*ihuenemátobou*), which could refer to the experience of the descent of the *ichéiricou*.

Although Breton classifies the beliefs and offerings to the *ichéiricou* as idolatrous,[63] he downplays the role of material idols and dissociates it from these beliefs. He only mentions such figures in the entry for *mánlira*,[64] the incorruptible wood of guaiacum which was employed for the pillars of the communal houses and for children's toys, which also received that name. However, the friar mentions that he found, hidden under some rocks and abandoned, beautiful and well-preserved figures made of this wood, which, out of superstition, the naturals did not dare look at anymore.

Breton also downplays and trivializes the role of bodypainting, remarking on the humorous and grotesque figures and patterns youngsters painted on their faces with *cóina,* the soot produced by the gum of the *elemi tree.*[65] He treats in a similar fashion the *imitóni:* or disfiguring patterns, which women painted on men, preparing them for some festivities.[66]

Although he recognizes a relation between *achdóüata,* the darkening with *chdóüa* (*Genipa americana*), and *kachoulocdtium,*[67] the sucking of disease by the *boyáicou,* he also notes that this tincture was employed as a preventive cure against *yaya,* that is, syphilis or yaws.[68] Similarly, he acknowledges that the *tibíen,* the red tincture of *bichet* (*Bixa orellana*)[69] macerated in palm oil, and mixed with coal from sandal wood, with which the men normally painted their body, was a remedy (*ibíen*). This tincture, he notes, closes the pores of their skin, protecting it from seawater, mosquitoes, and parasites. However, elsewhere he also notes the Callinago would mix with this tincture the juice of a variety of *taya,* a Caribbean cabbage (*Andira* sp.),[70] to make *abiénronê,* a fascinating spell, which would draw or attract the sympathy of enemies, preserving the painted person from their attacks.[71]

Rochefort claims *mápoya* constantly tormented the Caribs. Out of fear and for protection, they carved on hard woods the figure with which these spirits had appeared to them and wore them on their necks as amulets.[72] Breton fails to mention these figures. However, he is willing to acknowledge the curative potential of diverse adornments (*íari* or *énega*), remedies, amulets, or relics, with which the Callinago adorned themselves. Breton lists a variety of them—the most important and valuable were made of *calloucoulli* or *caracoli,* an alloy of gold, silver, and copper (which European smiths never managed to replicate successfully) which was used as nose ornament or as a pendant in the form of either a circle or a crescent moon, encased in hard woods.[73] Two kinds of green stones, presumably emeralds and jade, were highly treasured; they were employed against kidney stones, to 'alleviate women', and as a remedy for epilepsy.[74] These adornments could be related with a form of sorcery *línharonê* (from *iára* bewitching).[75]

Breton takes particular care to camouflage entries referring to sorcery distinguishing these practices from solemn festivities. For instance, illustrating Callinago approaches to liberal arts or medicine, he describes the *ébene* (a lancet made with agouti teeth and bone), employed to scratch the skin, to drain and cleanse blisters and sores, or to expose the skin to 'remedies'. The entry *Ebénnêtou* presents no translation but refers to *chebéneboüi tíatina*: she cast a spell on me. Under this entry he defines *Ebénnêtou* as sorceress. He continues with the *ébeti* medicine—neckless small elongated gourds filled with the skin of *caicouchi* (jaguar),[76] the meat of *anánna* (hawk),[77] claws of birds of prey, or similar things, which the Callinago took everywhere, specially to feasts. The Callinago employed the meat of *anánna* to make themselves brave. In other cases, Breton was not certain but suspected that they carried these substances out of superstition, to preserve themselves from being cursed or to make themselves brave.[78] During the *Eléoütac*,[79] a feast for married men with children, the participants would expose themselves (*chibanatoumain* or *cheboüitoúmain*), which the friar compares to a military initiation. Their captain acting as a *ioubénene*, a godfather, would break the head of a bird of prey, caught and fed for this purpose, spilling the bird's blood over the initiate's head. Then he would scratch their skin with the *ébene*, and rub them with blood and chilies. Afterwards, they would rub their wounds with the powder of jaguar skin. Children were taken to bed without eating; they were placed in a particular position which they were supposed to hold through the night, under the risk of deformation. Adults would get extremely drunk forcing themselves to ingest large amounts of *ouicou*. The powder of jaguar skin, comments Breton, was very effective drying the wounds, preventing inflammation and the formation of abscesses.

Breton sought to open the minds of missionaries towards the Callinago, demystifying many of those characteristics that had stigmatized them. Simultaneously, he sought to establish a place for missionaries in Callinago society. The *boyáicou* were rivals to be disenfranchised, and he sought to isolate and differentiate their practices from those of the wider community, in particular those of Callinago leaders, whose friendship he successfully courted. He attempted to trivialize and to find European analogues for the solemn ceremonies conducted by Callinago leaders, regarding them as superstition, actively camouflaging their connections with sorcery. On the other hand, Breton was also attentive to the resources the islands could offer and sought to naturalize Callinago sorcery and medicine.

3.3 Analysis: The Spirits of the Encounter

The Callinago were communities composed of at least two ethnicities, and their language was closely related with modern-day Arawak and with Taino. Nevertheless, they characterized themselves and their allies—the Galibi nations of the continent—as the rightful people, to be distinguished from the *other* peoples, the *Arouages* and the Christians. Their ethnical affiliations included non-human beings, such as the *ichéiricou*.

Like Urayoan, who conducted 'experiments' to determine the nature of the Spaniards bodies, the Callinago had a clear inclination towards empiricism, refusing to acknowledge as real what they could not see or use. These inclinations and experiments question the claims that Amerindian ontologies ascribing to multinaturalist metaphysics are diametrically opposed to naturalistic multiculturalism and that:

> The shaman walking through the forest does not question whether spirits exist (a multicultural question) he only wants to actualize his relations with them.[80]

The shaman does not question the existence of shamanic beings in the forest, because he either experiences them or at least can expect them to be there, they are part of reality. Urayoan and the Callinago were experiencing a multicultural reality, and they sought to ascertain how the reality Europeans brought with them corresponded to their own ontologies.

The Callinago considered the body to be malleable and permeable, liable to acquire and transmit to the progeny physical and psychological qualities of substances consumed: the desirable bravery of a valiant enemy or of a hawk, the undesirable eyes of a manatee, or the stupidity of a turtle. This understanding was at the base of some of their food prohibitions and also explains the rigour of their fasting practices. In a similar fashion, body painting served as a means to protect the skin, ward off disease, or draw the sympathy of adversaries seducing them. On the other hand, the Callinago also fostered the permeability of their body slashing their skin and exposing it to substances such as the blood of the hawk,[81] aiming to acquire qualities such as bravery. These procedures had material effectiveness, as attested by the cicatrising power of jaguar skin powder.

While recognizing the power of Callinago natural knowledge, Breton sought to purify the knowledge and practices from those of the *boyáicou*. Like Pané before him, Breton reduced the *boyáicou's* practices to scams and mockery of Christian rituals, deriving their power from delusion and suggestion. That is, Breton attributed to the *boyáicou* the employment of means which are at the basis of Christian faith, the power of words and symbolic actions over the material body. The description of the 'scams' portrayed by these friars and their interpretation are replicated by Lévi-Strauss (1958), as he analyzes the testimony of an Amerindian Christian convert who reluctantly becomes quack shaman, to conclude that the power of shamanic procedures depends solely on their symbolic efficacy.

Despite remarking that the Callinago lacked a spiritual language, Breton translated various notions as souls and spirits. The notions he translated as souls *ioüánni*: denoted primarily the heart and the pulse, the term was also related with *Iaoüa* shadow and *Iaoüátêna* image. Some of these notions of 'soul' constituted a plurality that participated in an extended version of 'metempsychosis', in which there was a constant transmigration of 'souls' which was not limited to the animal kingdom but included even rocks. After death, some of the 'souls' remained in the bones, which were thus not considered inert. Others migrated either to the sky, to the forest, the seaside, or the sea, becoming respectively *ichéiricou, mápoyanum, oumécou*, and *couloúbi.*

Taino *opia, çemies*, and *behiques* or *buhities* find continuity in the Callinago *oupoyem* or *acansáncou*, *çemijn* or *ichéiricou*, and *boyé* or *boyáicou.* These notions are further expanded, categorized, and developed. The general categories of *acansáncou* or *oupoyem*, through which Bretón translates the general notion of spirits, referred to the agents that constituted and interpenetrated Callinago bodies. However, *oupoyem* was related with matter and with the *boyáicou*, the physicians-magicians. *Acansáncou* was related with understanding and science, with form, figure, and what he translated as soul, but also with shadow and image (see the following chapter and the **Second Part** for further development). Besides *ichéiricou,* these categories included also 'evil' spirits. Although Bretón claims to support his translation of the *Callinago* term for devil in their animadversion to *mápoya*, it is possible that the moral classification of good and evil spirits was borrowed or that it was a veil to cover practices of the general community and their leaders which would be disapproved by missionaries.

These entities were related with diseases—they were pathogenic. They required nourishment, grew, matured, and aged; had genders; and reproduced prolifically. These beings recognized an extended progeny with which they feasted, they also had social structures, coinciding with those of the humans, such as nations with which they cohabited. These entities could be perceivable after prolonged fasting and the consumption of tobacco, through fantasies (visions) or dreams. Although the friar neither understands nor indeed claims to experience these phenomena, the fact that the *boyáicou* could claim that the events occurred 'before everyone's eyes' and that the other participants confirmed these experiences or were enthralled by them implies that these phenomena were a commonly experienced reality, an evident fact.

Entities classified as 'evil spirits' were closely related with the forest (*mápoyanum*), the shore (*oumécou*), or the depths of the sea (*couloúbi*). These entities were perceivable through smell and caused disease and mishap, effectively controlling these spaces, but also provided for fishing and hunting. Thus, Caribs avoided annoying them and acknowledged them through effigies of their image, as well as various medicines or amulets worn in their neck.

The etymological relation between *boyáicou*, *oupoyem* as a general category and *mápoya* in particular, suggests that there was a degree of identification with these entities. *Boyáicou* underwent an apprenticeship, through which they inherited descendants of their master's *ichéiricou*, who developed a relation of exclusivity with them. Breton shows that there were female as well as male *boyáicou*, and suggests that there was a correspondence between their genders and that of their *ichéiricou*.

Ichéricou regulated the growth of cultivated staples like manioc and controlled the weather. They caused disease on enemies of their *boyáicou*, gave advice about healing through medicinal plants, combated *mápoyanum*, and assisted in the process of sucking the arrows of disease. They also gave advice regarding a variety of political matters. Although Callinago would recognize *ichéiricou* as creators of their world, they rejected the notion of a universal creation.

Zemes or *ichéiricou*, and other shamanic beings described by missionaries, allowed missionaries to translate and introduce their spiritual notions. Nevertheless, the Callinago were clearly inclined towards empirism, and these entities were not ethereal, but perceivable; they had bodies, as well as many of the properties of living and social beings, features which are incommensurable and incompatible with Christian understandings of spirits.

Notes

1. The original relation of Pané was lost; the only surviving text is found in Ulloa's translation to Italian of the diary of Ferdinand Columbus 1573. However, an abridged version of the text is found in the *Decades* of Peter Martyr d'Anghiera, first published in 1504. De las Casas also commented the text. Arróm presents an annotated translation to Spanish based on Ulloa's translation, with the texts of d'Anghiera, and Las Casas as Appendixes (Pané 1988b [1500], 11). All quotations to these three authors are my translations from Arróm's Spanish translation.
2. Arróm employs the modern Spanish spelling *cemies*, I will follow *zemes*, the English spelling featuring in the OED.
3. My translation from (Pané 1988a [1498], 21).
4. Pané 1988b [1500], 34–35.
5. Stevens Arroyo 1988. This interpretation is derived from the interpretation of a passage in Pané (Pané 1988b [1500], 53). In one of the disputes between the converts of Pané and an antagonist chief, the followers of the chief took advantage of an absence of the priest to break into the common house of the Christians, pulled down the cross, tore it to pieces, and interred the pieces in a field. On these grounds grew chili, which the friar thought had been planted to mock Christians. The Spaniards

imposed the death penalty on the Taino responsible for the 'vituperation'. Various authors have suggested that rather than vituperation, the actions were the result of a misinterpretation—the assailants assumed Christ to be equivalent to Yucahu Bagua and acted accordingly (Pané 1988b [1500], 81). However, this interpretation seems unlikely given the marked antagonism of the relations between the friar and the chief.

6. Joyce 1916; Rouse 1993, 13,118–19; Olsen 1974, 89–102.
7. Breukel 2013. In Japan, calcirudite has been traditionally associated with national prosperity and thereof worshiped. Modern geoengineering employs nodules of calcirudite to improve the soil. The nodules act as a substrate for ureolytic bacteria carrying carbonate diagenesis, a process whereby the nodules are partially dissolved, and precipitate again with additional carbonates. The process contributes to the fertilization of the soil and cements lime and sand soils, improving the texture of loose grounds (Yong et al. 2014, 449–456).
8. The crown could refer to the caul: the amniotic sac. For some Amerindian communities like the Gunadule, the patterns of the caul (*gurgin*) are a sign, which evidence a relation with particular shamanic beings, and which confer aptitudes for hunting, fishing, or shamanism (Fortis 2010).
9. Pané 1988b [1500], 41.
10. Assuming *buhitie* as an Italianised form, Arróm translates *behiques,* which is the term employed by Herrera. However, Oviedo also employs *buhitie* indicating that this spelling was also acknowledged by Spanish speakers. As will be seen in the following section, it is possible that both forms are correct, representing different conjugations.
11. Snuff containing tobacco and the powdered shell of *covo* sea-snail (*Strombus gigas*), which was blown through nose-pipes to inhale. Arróm suggests *cohoba* should contain *Anadanthera* spp. or some other strong psychotropic substance. However, no traces of such substance have been found to be associated with Taino archaeological remains (Oliver 2009). On the other hand, many forms of Amerindian shamanism rely exclusively on tobacco. Furthermore, many Amerindian shamanic traditions employing stronger psychotropics consider tobacco as a more important substance for their practice (Russell and Rahman 2015).
12. Pané 1988b [1500], 35–37.
13. Pané 1988b [1500], 42.
14. Pané 1988b [1500], 38–40.
15. Pané 1988b [1500], 31–33.

16. Fernández de Oviedo 1959 [1535], Part 2. Vol. 126–127.
17. Fernández de Oviedo 1959 [1535], Part 1, Vol. 1, 479.
18. Fernández de Oviedo 1959 [1535], Part 2, Vol. 1, 126–127.
19. Lévi-Strauss 1987; Latour 2009; Viveiros de Castro 1998.
20. Breton and Troiani 2001, 157. Troiani presents a translation to Spanish of the French original, and all quotations referred to as (Breton and Troiani 2001) are my translations to the English from this version. All those cited as (Breton 1665) are my translations directly from the French original.
21. Breton 1665.
22. Du Tertre 1954 [1654].
23. de Rochefort et al. 1965 [1665]. Tylor's main source for the Caribs.
24. Breton distinguishes female (f.) terms, but we will distinguish both female (f.) and male (m.) terms wherever possible.
25. Taylor 1954.
26. Pury and Lewis 2001, 194.
27. Breton and Troiani 2001, 87.
28. Breton 1665. My translation, in the introduction, no pagination.
29. Breton 1666, 158.
30. Breton states that verbs and names are frequently followed by their contracted form, which is the way in which they are modified by the pronoun or by conjugation. However, the termination -cou is frequently employed in plural forms.
31. Breton and Troiani 2001, 103.
32. Breton and Troiani 2001, 75. My translation.
33. Breton and Troiani 2001, 21.
34. Breton and Troiani 2001, 34, 35 & 159. According to Breton the b and the p were frequently exchanged (p. 161).
35. Breton and Troiani 2001, 162.
36. Breton and Troiani (2001, 39) comment that the Callinago pronounce the ç as the French ch, so it was probably pronounced scheemein or shaman. It is thus possible that this is a parallel origin for the word shaman.
37. Breton and Troiani 2001, 54.
38. Breton and Troiani 2001, 168.
39. Breton and Troiani 2001, 108.
40. Breton and Troiani 2001, 131.
41. de Rochefort et al. 1965 [1665], 279.

42. de Rochefort et al. 1965 [1665], 288.
43. De Rochefort uses both *maboya* and *mapoya,* noting that the b was pronounced as in German. Given that Breton (Breton and Troiani 2001, 73) reports that *Couloubíni* was deep water, it is possible that *Couloúbi* were *acánsancou* of the depths.
44. de Rochefort et al. 1965 [1665], 272–273.
45. Labat 1724, 166–167.
46. Du Tertre 1954 [1654], 404.
47. de Rochefort et al. 1965 [1665], 280.
48. de Rochefort et al. 1965 [1665], 280.
49. Breton and Troiani 2001, 21 26, 50, 73, 118. The word boyé also denominates fireflies (Breton and Troiani 2001, 35).
50. de Rochefort et al. 1965 [1665], 312–313.
51. Breton and Troiani 2001, 91.
52. Breton and Troiani 2001, 6.
53. Breton and Troiani 2001, 168. Goeje (1942, 272, 1928, 272) describes a similar procedure for training of Semeçi (shamans) amongst the Arawak of Guyana, after fasting on a diet of tobacco balls and tobacco water, and vomiting for various days, the apprentices inhale the fumes of the boiling juice of *takini* bark, visions appear and the *semeçi* interprets their meaning.
54. Breton and Troiani 2001, 109.
55. My translation from (Breton and Troiani 2001, 85). In the first part, 'Hurry… escape', the friar's translation is straightforward. Afterwards, he is more imaginative. A literal translation would read 'eat everything, all feet, all hair, all excrement, feast on everything/. However, *ouáttê*: excrement, leaves some room for interpretation. Under *ouáttê* the friar links the entry *ioüáti:* 'in me', the reading would be less scatologic, simply: 'in me all'. However, faeces and their transformations across the perspectives of different beings are recurrent themes in Amerindian mythologies: the excrements of the gods are gold for humans, the faeces of humans are ornament for sloth spirits (Lévi-Strauss 1988, 126).
56. de Rochefort et al. 1965 [1665], 327–329.
57. The *boyé,* or the phantasy of the friar, could also be playing with a double sense (Breton and Troiani 2001, 85), sexually threatening the celibate friar. The sexual connotation is veiled in the dictionary through the misplacement and conjugation of the entries: *oüatáloucabouicienrou*: she committed an adultery (p. 156) and *Oulemátêti oüatállounombuíkêti*:

committed adultery (p. 161), and the misplacement of the entry *Ouáttou* fire (p. 157). The misplacement could be concealing practices of sodomy. Note also that different spellings may have resulted from the conjugation of the word as in *chiouátte hóman huibónem* or *kibonem hioüatekê*: tie or hang our hammocks (p. 64).

58. Breton and Troiani 2001, 43.
59. *iégue* (Breton and Troiani 2001, 111).
60. de Rochefort et al. (1965 [1665], 303), amongst the Arawak Apurina, even moieties within the same community have different food taboos (Virtanen 2015).
61. Breton and Troiani 2001, 105.
62. Breton and Troiani 2001, 143.
63. Breton and Troiani 2001, 24.
64. Breton and Troiani 2001, 135.
65. *Bursera* sp. (Grandtner and Chevrette 2013, 81; Breton and Troiani 2001, 69).
66. Breton and Troiani 2001, 114.
67. Breton and Troiani 2001, 6.
68. Grandtner and Chevrette 2013, 262; Breton and Troiani 2001, 185.
69. Grandtner and Chevrette 2013, 65; Breton and Troiani 2001, 33.
70. Grandtner and Chevrette 2013, 32.
71. Breton and Troiani 2001, 24 & 174.
72. de Rochefort et al. 1965 [1665], 282.
73. Breton and Troiani 2001, 43 & 82; de Rochefort et al. 1965 [1665], 258.
74. Breton and Troiani 2001, 171.
75. Breton and Troiani 2001, 104.
76. *Panthera onca*.
77. *Mansfenix*. The description in Du Tertre (1954 [1654], 359–360) coincides with the red-tailed hawk (*Buteo jamaicencis*); he notes that the meat was dark but delicious.
78. Breton and Troiani 2001, 76.
79. Breton and Troiani 2001, 80.
80. Kohn 2015.
81. If we take into account the pivotal role of vultures and birds of prey in modern Carib shamanism (Jara 2002), we are on stronger grounds to question the 'secularity' of these rituals.

Bibliography

Breton, Raymond. 1665. *Dictionaire caraïbe-françois, Meslé de quantité de Remarques historiques pour l'esclaircissement de la Langue*. Auxerre: Gilles Bovqvet, Imprimeur ordinaire du Roy. Original edition, 1666.

———. 1666. *Dictionaire français-caraïbe*. Auxerre: Gilles Bovqvet, Imprimeur ordinaire du Roy. Original edition, 1666.

Breton, Raymond, and Duna Troiani. 2001. *El caribe insular del siglo XVII Tratado sobre la lengua y la cultura de los Callínago Traducción al español del Dictionnaire caraïbe-français de Raymond Breton*. París: CELIA-CNRS. Original edition, 1665.

Breukel, Tom. 2013. Threepointers on Trial: A biographical study of Amerindian ritual artefacts from the pre-Columbian Caribbean. Master's Thesis, Archaeology, University of Leiden.

de Rochefort, Charles. 1665. *The History of the Caribby-Islands: Viz. Barbados, St Christophers, St Vincents, Martinico, Dominico, Barbouthos, Monserrat, Mevis, Antego, &c. in All XXVIII. In Two Books. The First Containing the Natural; the Second, the Moral History of Those Islands. Illustrated with several pieces of sculpture, representing the most considerable rarities therein described. With a Caribbian-vocabulary*. Trans. John Davies. London: T. Dring and J. Starkey.

de Rochefort, Charles, Charles-César comte de Rochefort, César de Rochefort, Jean Baptiste Du Tertre, Louis de Poincy, Raymond Breton, and John Davies. 1965 [1665]. *The history of the Caribby-Islands: Viz. Barbados, St Christophers, St Vincents, Martinico, Dominico, Barbouthos, Monserrat, Mevis, Antego, & c. in all XXVIII. In two books. The first containing the natural; the second, the moral history of those islands. Illustrated with several pieces of sculpture, representing the most considerable rarities therein described. With a Caribbian-vocabulary*: JM [ie John Macocke?] for T. Dring and J. Starkey, London.

Du Tertre, Jean-Baptiste. 1954 [1654]. *Histoire générale des isles de S. Christophe, de la Guadeloupe, de la Martinique et autres dans l'Amérique*: chez Jacques Langlois: et Emmanuel Langlois.

Fernández de Oviedo, Gonzalo. 1959 [1535]. Historia general y natural de Indias. In *Biblioteca de Autores Españoles*. Madrid: Biblioteca de Autores Españoles. Original edition, 1935.

Fortis, Paolo. 2010. The birth of design: A Kuna theory of body and personhood. *Journal of the Royal Anthropological Institute* 16 (3): 480–495.

Goeje, Claudius Henricus. 1928. *The Arawak language of Guiana*.

———. 1942. De Inwijding tot Medicijnman bij de Arawakken (Guyana) in Tekst en Mythe. *Bijdragen tot de Taal-, Land- en Volkenkunde* 101 (1): 211–276.

Grandtner, M.M., and J. Chevrette. 2013. *Dictionary of trees, volume 2: South America: Nomenclature, taxonomy and ecology.* New York: Elsevier Science.

Jara, Fabiola. 2002. The meaning of nominal animal categories among the Caribs of the Guianas. *Anthropos* 97 (1): 117–126. https://doi.org/10.2307/40465620.

Joyce, Thomas Athol. 1916. *Central American and West Indian archaeology: Being an introduction to the archaeology of the states of Nicaragua, Costa Rica, Panama and the West Indies, Central American archaeology*. New York: G. P. Putnam's Sons.

Kohn, Eduardo. 2015. Anthropology of ontologies. *Annual Review of Anthropology* 44: 311–327.

Labat, Jean-Baptiste. 1724. *Nouveau voyage aux isles de l'Amérique contenant l'histoire naturelle de ces pays, l'origine, les moeurs, la religion et le gouvernement des habitans anciens et modernes, les guerres et les evénemens singuliers qui y sont arrivez pendant le long séjour que l'auteur y a fait, le commerce et les manufactures qui y sont établies et les moyens de les augmenter...[par le P. Labat].*

Latour, B. 2009. Perspectivism:'Type'or 'bomb'? *Anthropology Today* 25 (2): 1–2.

Lévi-Strauss, Claude. 1958. *Anthropologie structurale.* Paris: Plon.

Lévi-Strauss, C. 1987. Raza e historia. *Antropología Estructural II. Siglo XXI.* México: 304–340.

Lévi-Strauss, Claude. 1988. *The jealous potter.* Trans. B. Chorier. Chicago: University of Chicago Press.

Oliver, Jose R. 2009. *Caribbean archaeology and ethnohistory : Caciques and cemi idols: The web spun by Taino rulers between Hispaniola and Puerto Rico.* Tuscaloosa: University Alabama Press.

Olsen, Fred. 1974. *On the trail of the Arawaks.* Norman: University of Oklahoma Press.

Pané, Ramón Fray. 1988a [1498]. Relación de Fray Ramón acerca de las antigüedades de los indios, las cuales, con diligencia, como hombre que sabe la lengua de ellos, las ha recogido por mandato del Almirante. In *Relación acerca de las antigüedades de los indios: El primer tratado escrito en América*, ed. Jose Juan Arrom. Bógota: Siglo XXI Ediciones.

———. 1988b [1500]. Relación de Fray Ramón... In *Relación acerca de las antigüedades de los indios: El primer tratado escrito en América*, ed. Jose Juan Arrom. Bógota: Siglo XXI Ediciones.

Pury, Sybille de, and Marcella Lewis. 2001. *The language of the Callinago people: Father Breton's dictionnaire caraïbe-français (1665) compared with Garifuna.* Paris: Sup-Infor. http://www.sup-infor.com/ultimes/garifuna_a/garifuna_a-ina.htm.

Rouse, Irving. 1993. *The Tainos: Rise and decline of the people who greeted Columbus.* New Haven: Yale University Press.

Russell, Andrew, and Elizabeth Rahman. Eds. 2015. *The master plant: Tobacco in lowland South America.* London: Bloomsbury.

Stevens Arroyo, Antonio. 1988. *Cave of the Jagua: The mythological world of the Tainos.* Albuquerque: University of New Mexico Press.

Taylor, Douglas. 1954. A note on the Arawakan affiliation of Taino. *International Journal of American Linguistics* 20 (2): 152–154.

Virtanen, Pirjo Kristiina. 2015. Fatal substances: Apurinã's dangers, movement and kinship. *Indiana* 32: 85–103.

Viveiros de Castro, Eduardo. 1998. Cosmological deixis and Amerindian perspectivism. *Journal of the Royal Anthropological Institute* 4 (3): 469–488.

Yong, R.N., C.N. Mulligan, and M. Fukue. 2014. *Sustainable practices in geoenvironmental engineering.* 2nd ed. New York: CRC Press.

4

Syncretic Ontologies of the Microbial-Shamanic Beings

This chapter[1] examines the similarities between the notions of Amerindian shamanism and of microbiology; it suggests the common grounds shared by both ontologies and proposes the hypothesis that some shamanic beings might be microbes. The chapter starts reviewing the notion of shamanic beings, and how it appeared in early accounts and in recent ethnographies. Then the chapter exposes how these notions are paralleled by recent understandings of the workings of the microbiome. The chapter then questions the traditional position of anthropologists towards medical materialism, as well as some of the positions of science and technology studies towards microbes.

In the five hundred years that have passed since the Encounter, shamans have been stigmatized, ostracized, and persecuted by ecclesiastic and civil authorities, who emphasized the relation of their practices with sorcery or witchcraft. Shamans appropriated this role; fostering this image, they impersonated the colonizers' guilt and terror, acquiring power over them.[2] On the other hand, in the last century in the wake of anthropological and ethnobotanical interest, various countercultural movements promoted a romanticized image of shamans: of their understanding of reality,[3] and their idyllic relations with nature, giving rise to

C. E. Giraldo Herrera, *Microbes and Other Shamanic Beings*,
https://doi.org/10.1007/978-3-319-71318-2_4

neo-shamanic movements.[4] Shamans also appropriated this image and profited from forms of tourism associated with it, deriving considerable part of their sustenance and political power therefrom.[5] Both trends could affect the reputation of shamans, who are often considered as dubious characters even by their own communities and by other shamans.

Nonetheless, part of the reason shamans successfully adapted to these roles is that they were already part of their repertoire. These images do not detract from the fact that shamans are a particular kind of medical specialists found among various others in Central and South American communities. The realm of medical knowledge of shamans is associated with their concomitant functions in wildlife or more widely environmental management. They diagnose and treat a variety of diseases, many of which are categorized in Western biomedicine as infectious and contagious.[6] Although many of the notions of contemporary Amerindian shamans have been influenced by the advent of Christianity and of modern medicine, many of their notions display commonalities or continuities with those of the groups of the Encounter.

4.1 Shamanic Beings

Taino and Callinago notions of *acansáncou* find a close parallel in the notions of many contemporary Amerindian communities, such as the Gunadule, the Tukano, the Arawak, and the Yanomami.[7] I will examine the Yanomami notions of *utupë and xapiripë,* which are beautifully exposed by Yanomami shaman Davi Kopenawa and Bruce Albert,[8] allowing the elucidation of many aspects of these concepts. *Utupë* has been translated 'as image, vital principle, true interiority or essence'.[9] Strictly speaking, *utupë* refer to multitudinous entities, with their own personhood and intentionality that inhabit, mirror, and holographically constitute each being of the environment: humans and animals, plants, soils, bodies of water, rocks, and even tools.

For many contemporary shamans, the complexity and interactions of the multiplicity of image/beings constituting the body reflect those of the wider environment. Following environmental disturbances, these entities can predate upon one another, action which is correlated with disease. In this way for instance, Davi Kopenawa describes a particular kind of *utupë* denominated *xapiri pë*. *Xapiri pë* are the owners and protectors of wildlife; they predate upon the predators of their protégées, and upon the transgressors of places under their protection. The differences between humans and animals are negligible compared to those between the former two and *xapiri*. They are the *utupë*, not of modern animals but of their ancestors.[10] *Xapiri* are tiny, like sparkling specks of dust, yet they are magnificent, luminous beings painted in bright red colour, black enveloping patterned designs, and wonderfully adorned with bright, white shining headdresses, colourful armbands, and waist plumes.[11] *Xapiri* replicate like mirror images, thereby escaping death.[12]

Xapiri are some of the pathogenic agents with which Yanomami shamans deal, and through whose agency they cure. In fact, the word also designates shamans themselves, who on account of their high permeability to these entities have a corpse-like rotting quality and are understood to be on par with these beings.[13] However, *xapiri* are most effective curing diseases caused by beings they have always known. When they confront the *xawarar*i—beings brought by white people, who cause the various *xawara*—contagious infectious diseases of foreign origin[14]—fevers debilitate them and the *xawarari* devour *xapiri* as well.[15] Only a few, like the eel *xapiri*, know how to strike certain *xawarari*.[16]

The power of Yanomami shamans depends on the number, the diversity, and the particular *xapiri* they host or those they can summon to their assistance. *Xapiri* come to inhabit what they regard as a communal house in the chest of shamans. Shamans have to work continually to make themselves habitable for particular *xapiri*, undergoing regimes of fasting, dieting, abstinence, and purges.[17] These conditions prevent odours of rotten food, in particular game, smoke, copulation, excrement, or honey flowers, which some *xapiri* find disagreeable,[18] promoting an environment in which *xapiri* may build the house and orderly populate it, paralleling the ecological succession of the forest. Powerful shamans call their *xapiri* singing their songs and pass them to initiates along with their breath while they blow *yakoana*[19] powder into the latter's noses.

The process starts by summoning transitory *xapiri*, who, through ephemeral dances, prepare the ground and build the house, allowing for the establishment of other *xapiri,* which will settle permanently. The first to be called are the *xapiri* of clearing birds, such as those of the curassows, which like the latter find nourishment sweeping and cleaning the forest floors. These *xapiri* are followed by the *xapiri* of the leaves and roots of lianas and trees, which clean the dirt. In turn, the latter are followed by those of white rocks and termite mounds, which cover the grounds with their feathers. Through this process senior shamans must hunt and keep at bay the *xapiri* of snails, as well as household *xapiri* associated with ashes, firebrand, cotton hammocks, carrying baskets, and pubic aprons. These *xapiri* are considered too weak and thus unable to cure.[20] Moreover, forest *xapiri* loath them and would not approach an initiate if they had presented their dances to him. With the ground cleared and protected, the *xapiri* of howler monkeys, of the sky, and the storms, come from afar, carrying with great effort the communal house already made, coordinating their activity with their songs. Once in place, the *xapiri* of spider monkeys and sloths secure them with ropes and nails. Spider *xapiri* then cover the roof with leaves that have been carried by the *xapiri* of the great anteater.

Once the house has been built, senior shamans send their *xapiri* messengers to invite important *xapiri* to feast. If the house is well built and clean, many *xapiri* will gather, presenting their dances, which the initiate experiences as terrifying visions, which may hurt him, inflicting deep wounds on his body. The initiate must learn the songs to be able to reply and make them welcome.[21] After the feast, some of these forest *xapiri* will settle in the initiate's *xapiri* house, hanging their hammocks, and making their own homes. *Xapiri* do not readily mix with one another, so there are various distinct homes in a communal house: one for the *xapiri* of bees, and another for those of the anaconda, the jaguar, and the puma. Some like the tapir *xapiri* will share with their in-law-sons: the *xapiri* of the falcon and the *xoapema* bird, or the *xapiri* of the storm who cohabit with those of thunder. Others like those of the toad live alone. The houses of evil *xapiri* like the *koimiri* bird of prey, the anaconda, the jaguar, and so on have to be placed in isolation, on the top of the *xapiri* house, because these are fierce and dangerous beings who hunger for human flesh.

The *xapiri* houses of the initiates are small and cannot accommodate many different *xapiri.* Shamans have to work progressively to enlarge them. This process is regulated by an increasingly diverse diet, which coincides with the dietary requirements and preferences of the *xapiri* involved; at first, the initiate must fast; after five days, he can start to eat honey; and still later fruits, dully cooked manioc, and white fish. Finally, after a long time, when the shaman has become accomplished and the jaguar or puma *xapiri* have performed their dances, they can finally eat game meat.[22] The building and maintenance of the *xapiri* house is a continuous process. If shamans neglect their *xapiri* houses, their own health deteriorates, and they may fall sick and perish. Each *xapiri* can be employed to cure certain diseases; however, they may be also employed to attack individual persons, for instance, collecting the *utupë* that person leaves in the ground on the tracks of their footsteps and calling *xapiri* against them.

4.2 Commensurable Aspects of Microbes with Shamanic Beings

The notions of *acansáncou, zeme, xapiri,* those masters of game and other shamanic beings might still seem alien to Western sciences and more akin to religious notions. Likewise, the practices of shamans, both of the present and the past, might seem superstitious, ethereal, and mystical. Yet, I want to argue that they are highly coherent with microbiology, in particular with its recent developments.

Through microbiology[23] we have come to understand that entities—normally invisible to the naked eye—inhabit the air, the water, and the earth, affecting their properties and dynamics.[24] Moreover, we, like other plants and animals, are ourselves microbiota, complex ecological systems in constant change.[25] We have recognized that we are made of cells, but more importantly that our mammalian cells are outnumbered by a highly diverse community of microbes, mostly bacteria. They make our bodies operational, foster our development, and mediate our ability to interact with the environment.

Microbes or microorganisms constitute an extremely wide and diverse category, including eukaryotes, prokaryotes, and even viruses. Roughly, this category comprises all the organisms with dimensions a thousand times smaller than millimetres.

We—all animals, plants, and fungi—are eukaryotes (organisms with cells) that have organelles (complex structures which perform differentiated functions). Amidst eukaryotic microorganisms we find some which can be readily classified as plants like algae, animals like the crustacean daphnia, and fungi like yeasts. However, amidst microbial eukaryotes we also encounter beings which are more difficult to classify, having properties of both plants and animals, and also symbiotic associations of these organisms, such as microscopic lichens, in which algae collaborate with fungi. Some eukaryotes like amoebas are composed of a single cell, others like *Volvox* are constituted by many, still others undergo through free living single cell and multicellular stages of development.

Prokaryotes are organisms without organelles. They are constituted solely by a cell wall or a membrane, containing various plasmids—loose chains of DNA, each codifying only few genes—and a dispersed biochemical arsenal. Despite their generally smaller size and apparent simplicity they constitute not one but two domains: archaea and bacteria. Each of these domains is larger and more diverse than all eukaryotes together. Moreover, these two domains are more distantly related to each other than to eukaryotic organisms. Prokaryotes are found in the most diverse environments, adapted to the widest ranges of conditions: some can be found at extremely high temperatures, in the high-pressure environments of deep underwater volcanic vents, others can be found hundreds of meters under the crust of the earth, deep in the ice, in soils, rivers, at sea, in clouds, as well as in the upper layers of atmosphere.[26] Some feed on chemical reactions. Others carry photosynthesis, feeding from the sun, like plants. Still others feed on death bodies. Many are predators. A few are parasites, causing diseases. Many more are mutualistic collaborators in symbiotic relations, providing crucial nutrients necessary for plant and animal growth.

Viruses are even tinier and simpler organisms constituted barely by a protein shell and some genetic material either as DNA or as RNA. They are generally not considered living organisms, because they do not reproduce by themselves, but require cells to carry out this work. Viruses are frequently parasites causing diseases. However, viruses also play crucial, often positive, roles in the development and in the evolution of all organisms.[27]

In many ways we—all eukaryotes—are descendants of some prokaryotes. The organelles of the eukaryotic cells evolved from the cooperation of particular archaea and bacteria.[28] On the other hand, at some stage of our evolution, we were unicellular eukaryotes. Moreover, due to their capacity to carry asexual modes of reproduction, bacteria and many single-cell eukaryotes can generate replicas of themselves. Hence, although vulnerable as individual cells, they are practically immortal as entities. Some of them may be rightfully called our living ancestors, the living ancestors of all fungi, plants, and animals.

Each microbe has particular needs and produces particular wastes. Under appropriate circumstances, they reproduce and may proliferate. Most also have at their disposal biochemical means to fend for themselves, to attack or to appease others, to modify the physiology and behaviour of their hosts,[29] or to influence the dynamics of the wider environment.[30]

Many of them associate with their sibling replicas, constituting colonies which act in a coordinated fashion. Moreover, both prokaryotes and eukaryotes have been shown to communicate through chemical and luminous signals,[31] and to associate not only with other members of their colony but also across species, collectively sensing the conditions of the environment and coordinating their behaviour to produce massive effects for their benefit.[32] Cooperating and competing on the basis of communal information, microbes collectively take advantage of the resources available in the environment, face hostilities, or shape their surrounding environment.[33]

For instance, facing scarcity or otherwise hostile conditions, they may collectively go into dormant states. Alternatively, they may cause propitiate air flows, facilitating their dispersal through the environment.[34] Microbes are known to produce proteins that catalyse water crystallization, inducing the precipitation of rain or snow, playing an active role in the water cycle of environments like the Amazon rainforest.[35] In a rich environment, they may reproduce and proliferate in an orchestrated manner. Yet, in similar circumstances, some microorganisms may opt to build biofilms, large protein, or lipid networks, which allow them to stick to surfaces, protecting themselves from hostile conditions, but also allowing them to travel or to transport resources.[36]

Microbial communication also involves the horizontal transmission of genes between members of the same 'species', also across species, and even across domains: between archaea, bacteria, and eukaryotes. The horizontal transfer of genes allows microorganisms to share solutions to the particular challenges of a given environment.[37]

However, these various processes of communication and cooperation are not exempt of self-interested behaviours—there is also treachery and delusion. Some microbes in a population may produce signals, which are inconsistent with their subsequent behaviour, allowing them to exploit collectively induced conditions. Exhibiting the capacity to communicate, and to cooperate or to act individualistically, microbes can be considered as highly social,[38] both amongst themselves and in their complex and intricate interactions with their environments, be it soils, clouds, bodies of water, or bodies of plants and animals. Through these various processes, each contributes in one way or another to the general state of these environments, constituting highly specific, diverse, and dynamic ecological communities or microbiota, with characteristic, though also dynamic, collective genomes or microbiomes.[39]

Microorganisms, mostly bacteria, comprise more than half of the cells in the human body.[40] Furthermore, challenging Western notions of stable personhood and identity defined by the inheritance of DNA, microbes provide for roughly 98% of our working DNA.[41] Bacteria are found in each and every one of the surfaces of our bodies: in the skin, in our mouths, along the respiratory tract, in our reproductive organs, but they are most numerous and varied along the digestive track.[42]

Exposure to the environment plays a primary role in the constitution of our microbiota, while our 'human' genetic makeup appears to be subsidiary, affecting microbiota mainly through general features of metabolism.[43] The composition of the various microbiota in the surfaces of our bodies depends on the history of our interactions. Exposure to microbes starts early in life—bacterial communities from the mother's vagina and faeces colonize the skin, oral cavity, and the guts of the neonate during delivery. The process of establishment of these communities continues during the first year of life, shaping immune response and future adult microbiota, which roughly stabilize thereafter.[44]

Being highly exposed, our skin is the most dynamic of our microbiotas; nonetheless, it retains particular signatures, corresponding to our habits, to the substances to which we are exposed, and to the frequency of our interactions with other organisms.[45] The microbiota of the digestive tract is relatively more stable, depending on a history of exposure to particular bacteria, their relation with one another, and with incoming food and its bacteria, modified by changes in our dietary habits.[46]

Through our complex mutual interactions, microorganisms play a crucial role in our workings, both in health and disease.[47] Microbes project the body onto the environment and the environment into the body; nonetheless, they also demonstrate a high degree of autonomy. For instance, *Vibrio cholera,* frequently regarded as an obligate parasite, can survive in the environment under the right conditions.[48] Some microorganisms, frequently innocuous and even playing crucial roles in the microbiota of other organisms, produce toxins or predate upon our cells and on other of our inhabitants.[49] Others alter the behaviour of local communities, altering their stability.[50] They are therefore pathogenic, affecting our behaviour and transforming our bodies and behaviour.[51] For instance, the zoonotic protozoa *Leishmania braziliensis* produces cutaneous lesions medically described as tapir's nose,[52] which matches the metamorphoses described in ethnographical accounts of the counter-predation by masters of game.[53] These organisms leave traces of their presence in our immune system, and even in the human genome.[54]

Our interactions with microbes involve complex and diffused paths of interaction and perception. The vertebrate immune system has developed the capacity to recognize self- and non-self-molecules flowing through the body, developing specific immunities and acquiring a memory of novel encounters and successful responses.[55] This process is tightly connected with behavioural responses to such a degree that through evolution, its means of communication have been 'hijacked' by the neural system to deal with stress situations.[56] Moreover, these channels of mutual communication between microbial guest and animal host cells are not necessarily controlled by the latter.[57] Many bacteria produce the neurotransmitters and neuromodulators that mediate the actions of the system, for example, *Lactobacillus* produces acetylcholine and gamma-aminobutyric acid (GABA); *Bacillus* produces dopamine and norepinephrine; *Candida* and *Escherichia* produce serotonin (5HT). These and other microbes have the capacity to affect neuroendocrine function, our moods, and our emotions.[58] They moreover play a crucial role on its development. The serotonergic system, which plays a crucial role in emotion, and is one of the main targets of many of the psychotropic substances employed by shamans, depends for its development on our microbiota.[59]

Moreover, the brain and the gut microbiota are tightly connected through the vagus nerve, a bidirectional channel of communication, allowing the mutual influence. For example, facing rich food resources, some microbiota have been shown to signal the brain via the vagus nerve to stimulate parasympathetic activity augmenting food intake and increasing the storage in fat. This mechanism is adaptive for foragers in environments with scarce resources. However, it may lead to obesity in rich environments.[60] The microbiome and the immune system, in its association with the endocrine and the nervous system, provide multiple two-way channels of action-communication with microbes, with expressions through behaviour, emotions, and perhaps also dreams.

Microbes also mediate in our relations with our surrounding world. As by-products of their metabolism or as communicative signals,[61] microbes produce compounds with easily recognizable odours and tastes, sometimes revolting, others cherished, including the characteristic taste of liquors, wines and beers, of cheese and yogurt, but also coffee, chocolate, tea, and tobacco.[62] They also produce the smells that identify us personally and facilitate or interfere in our relations with others, humans and non-humans.[63]

Microbial ecology offers a view of the world highly coherent with shamanic ontologies. It portrays our bodies and environment as constituted by a dense and fluid population of highly social entities, which can affect our health and behaviour and play an important role on the environmental balance. Like Taino zemes, Callinago *acansáncou* or Yanomami *utupë* and *xapiri-pë* microorganisms are normally invisible, minuscule, multitudinous, and powerful. They are found everywhere, constituting and influencing the behaviour of our bodies and the environment, nourishing soils, plants, and animals. Like their Amerindian counterparts, microbes replicate overcoming death and some of them can be considered as our living ancestors. Moreover, they are also highly social, recognize progeny or sibling replicas forming colonies, and collaborate to produce massive effects, influencing the weather or causing disease. Thus, from a microbiological perspective, Amerindian alimentary and hunting taboos during pregnancy, hybrid notions of personhood, the constitution of the body through continuous contact with the environment, and the risks involved in hunting activities[64] are not unjustified and merit careful experimental examination.

Dealing with microbes, some of which are pathogens, seems by definition to be directly related with the shaman's trade. The permeability of shamans to pathogenic agents could be understood as openness to systemic infection; becoming infected, the shaman could identify the pathogen in himself and assist the patient's diagnosis or the assessment of zoonotic microbes in the environment. Permeability to microbes, however, does not necessarily imply that shamans are continuously sick. Provided the effectiveness of other immunity responses or the prevalence of mutualistic interactions in the host-pathogen relations, systemic infections can be medical conditions without developing into diseases.

Microbial ecology corresponds well with the medical specialty of shamans, whose knowledge is derived from and closely associated with wildlife management and the treatment of hunting-related (i.e. zoonotic) diseases. Assimilating shamanic beings to microorganisms does not imply a reductionist view so long as we remain aware of the complexity of the microbial worlds, and the potential insights offered by the subjective approach of shamans. As we will develop further on the **Second Part,** our sensorium, our organism, and its interaction with a wide variety of guests provide multiple potential modalities to explore our subjective interactions with the microbial world that constitutes us and the environment. Moreover, it is not necessary to relinquish symbolism but perhaps we should consider the possibility that symbols and the 'mind' or 'soul' might not be immaterial, abstract, or contained in neuronal vats but flow more freely and be widely distributed through the body and the environment.

4.3 Social Science and Its Recalcitrance to Microbiology

> Powell admired the Kaibab Paiaute as individuals,
> but regarded their culture as savage.
> Because … no society could claim to be advanced in which
> 'coughs are caused by invisible winged insects,
> rheumatism by flesh-eating bugs to small to be seen,
> and the toothache by invisible worms'.[65]

In anthropology we often rest assured that modern Western notions of germs or microbes have a historical and cultural context, which cannot be extrapolated beyond the time and society marked by Pasteur, as could be inferred from Douglas or Latour.[66]

4.3.1 Anthropology and Medical Materialism

'Purity and Danger' exposes the troubling and apparently antagonistic but coalescent notions behind the sacred in ethnographic and classical accounts. Among various 'primitive' peoples, what is sacred simultaneously connotes defilement, filthiness, and often contagiousness. This ambiguity is also found in the Latin root of sacred (săcer), the Greek for holy (ἱερός *hieros* and ἅγιος *hagios*), and the Hebrew *k-d-sh.*[67] The sacred is a point were magic, religion, and medical hygiene (science) meet in a most paradoxical manner, which has been troubling anthropology even before the discipline became independent from its theological missionary roots.[68] Nineteenth-century Eurocentric scholarship drew an ethical abyss between the materialistic but allegedly 'irrational' and profane belief in magic and the modern Protestant Christian enterprise. In an effort to salvage morality, Douglas sets out to bridge the divide and reunite magic with religion, at the expense of savage and pre-modern science. Vanquishing the prosaic shadows of heathen profanity, Douglas attempts to rid us of medical materialism,[69] that primitive sign of modernity.

In a most paradoxical way, she argues that the theory of pathogenesis and consequently contagion is only a very modern notion, which produced a revolution in medicine in the nineteenth century. For her, even in the West, the microscopic world is still supernatural to the layman and ultimately hygiene is limited to the control of filth and disorder; thus, it is a classificatory matter. Hence, it would have been unthinkable only two hundred years ago and is still foreign to any context not so evidently influenced by scientific notions of microscopic germs. Douglas, and anthropologist in her wake, exclude non-Western thought from science, secluding them to the symbolic realm. Anthropologists have been less concerned with distinguishing non-Western thought from religion, another Western category, which has devaluated and reduced to the realm of belief. In this way, anthropology contributes to disowning non-Europeans of their grip on reality, neglecting that science is now as was then the hard currency.

Douglas thesis depends on the assumption that there can be 'one range of dangers possible in any given universe',[70] and, moreover, that the range of dangers of a particular universe should somehow fit into a single logic. If in the twelfth century Maimonides could understand and explain physiologically all alimentary interdictions in Leviticus except for that of the swine, it is not because he had doubts in particular, but because there was no physiological reasoning behind the formulation of any of these rules.[71] If, on the other hand, Brahmans are stricter with hygiene than most Britons are, she does not consider that their line of work might require higher hygiene; instead, they must be acting symbolically.

Nonetheless, 'dangers' are specific possible outcomes of relations in a particular environment at a particular time; they depend on the activities and intentions of the organisms populating this environment, and this includes non-human entities. Frostbite is not an issue in the Colombian mangroves, nor is yellow fever an issue where the females of *Aedes aegipty* mosquitoes are not feeding.

Part of the problem is that the environment at stake is also a forest within, which, as we have seen, is only beginning to demonstrate its complexity to microbiologists. Echoing Brahmans, for whom, Douglas reminds us, the cow is not only a deity but in it reside millions of deities,[72] most of the cells of our body as well as those of the cow are not strictly mammalian but microbial. The developments in microbiological research and modern understanding of the body are, as Latour has shown, dependent to a great degree on the development of sterilization protocols, in other words rules of hygiene, which might seem no less absurd than those of Brahmans. Furthermore, in some instances, the procedures of Western science borrowed and misappropriated from non-Western practices.

Examples of such borrowing and misappropriation were the revolutionary descriptions of smallpox variolation by Jenner (1801), and his subsequent development of vaccines. By the time of Jenner's observations, variolation had been part of British folk medicine for over a century. The practice had been introduced from Middle Eastern traditional medicine. These practices had been further refined in India and Bangladesh where they were associated with the 'cult to a goddess'. Under British Colonial rule, these practices were forbidden and were replaced by comparatively cruder vaccination techniques, with catastrophic consequences for the population.[73] In the meantime, Jenner's scientific 'discovery' laid the basis for the contemporary understanding of immunology, which was crucial for the development of microbiology.

4.3.2 Historicity of Microbes

And yet, after a lengthy discussion on the development of the pasteurization process, Latour answers to the exasperated question: 'Did the ferments (or the microbes) exist before Pasteur made them up?'[74] And, he answers categorically: 'No!' The reason, he claims, is that we have to acknowledge that microorganisms have their own historicity.[75] Indeed, so long as this statement is not overgeneralized, it holds: Pasteur, and some particular strains of *Lactobacillus* involved in the creation of modern pasteurization, changed in the process. Pasteur and the *Lactobacillus* were made up by the process:

> Fermentation has experienced other lives before 1858 and elsewhere, but its new *concrescence**, to use another term of Whitehead's, is a unique, dated localized life offered by Pasteur – himself transformed by his second great discovery and his laboratory.[76]

What is implied is not that the *Lactobacillus* did not exist in an absolute sense, which would be equivalent to saying that there were no people, no persons or things in America before Columbus. It is rather that people who inhabited given locations in what today is America, at those times would not denominate or group themselves as Americans or Amerinds, nor could they be classified as such by foreign onlookers. Their 'identity' would not take that shape for many generations to come; their notions of personhood and humanity were radically different. Someone, something partially related to them existed, but not Indians or the Lactobacteria strains of Pasteur's laboratory.

Admittedly, the practices and instruments through which we approach and understand these beings (peoples and bacteria) vary; with these changes, our relations with them also vary, transforming us and them. The bacilli strains, which have been bred after Pasteur, were probably not found in the wild before his experiments, neither were they believed to be associated with yoghurt fermentation. Conversely, without these experiments Pasteur would not have become the figure he is recognized to be. Both were the product of an encounter in a scientific historical context as much as Columbus and the 'Indians', which does not mean that there had not been other such encounters before. In this limited sense, Latour's view is perhaps acceptable. However, if it is generalized to microbes at large, his claim does little justice either to microbes or scientists.

Trying to widen his scope, Latour recounts the dispute between Pouchet (*Spontaneous Generation*) and Pasteur (*Germ Theory*); he argues that at stake in this debate was the existence of either a life-force or microorganisms.[77] Reading between the lines of his portrayal, it would appear that the existence of microbes was already taken into account by both sides:

> Spontaneous generation was a very important phenomenon in a Europe devoid of refrigerators and ways of preserving food, a phenomenon anyone could easily reproduce in his own kitchen, an *undisputed phenomenon made more credible by the dissemination of the microscope*.[78]

Was life-force visible under the microscope? Why did it make spontaneous generation credible? What was arising spontaneously? Pouchet does not mention any observation of life-force under the microscope; instead, he presents the succession of microorganisms through observations carried out day-after-day in liquids or macerated solids and also in the mammalian uterus. Pouchet's lists of spontaneously generated creatures included a wide variety of organisms ranging from mammal ovaries, through other animals and plants to *Infusoria animalcules*, that is, *microzoaires, microphyta,* and *protozoaires*: '*Amiba, Monas, Trachelius, Bacterium, Vibrio, Spirillum*'.[79] The *Infusoria animalcules* and its sub-classifications have long become obsolete or have changed drastically; nevertheless, the names of its members and their corresponding descriptions should ring some bells; they are still the subjects of study of microbiologists, even though our conceptions of them have change drastically in the last hundred years. Furthermore, in a work written at the height of the dispute, Pouchet states:

> If our century clears to make justice to those ridicule hypotheses of panspermia, which impregnate all parcels of creation with germs and living organisms, we still have to recognize that if these untraceable metaphysical germs are nothing but a ridicule fiction, there exist nonetheless in the atmosphere, which appears so transparent and so pure some microzoaries that fly here and there…
>
> Man, himself, despite his pride, does not imagine what invisible population devours him incessantly until often sometimes ends up killing him. We discover today, in his intestine, masses of *Vibrions*, truly unperceivable eels. The mouth is perpetually inhabitated by myriads of animalcules in the tartar, which corrodes our teeth, and is nothing but a microscopic ossuary formed by the incrustations of their calcareous skeletons.[80]

In other words, what was at stake was not the existence of microorganisms, that they were alive and eating even within the human body, nor even that they could be airborne. Both Pasteur and Pouchet acknowledged these points, basing their views on the natural histories of microscopic organisms.[81] These accounts had been accumulating for more than a century and drew on micro-paleontological evidence which allowed the existence of these beings to be extrapolated into the distant past. The debate was limited to some of the attributes of life at large, specifically, the means through which living beings, including but not only microorganisms, originated: Pouchet was trying to prove through microorganisms that life could be generated *de novo*, while Pasteur, like Ehrenberg,[82] van Leeuwenhoek,[83] and Fracastoro,[84] argued that a fully formed microorganism had to be present or to be carried by air or water and be sowed into the medium to start reproducing and multiplying in it. In other words, and by analogy, what was discussed was not whether the platypus exists, but whether it is viviparous or oviparous. Had Pouchet triumphed, yoghurt would perhaps not be such a standardized product, and microorganisms would not have stopped sprouting pretty much out of everything, as they still do.

Backward causation and retrofitting are still problematic issues; should we hold reality hostage to the records of the history of mainstream ideas in Western science? The problem if we take Latour too seriously is that not only do microbes cease to exist before 1864, mammals, humans included, are also banished from the picture, or at least we become entirely different (spontaneously generated) entities. But, is identity ever so fixed? Perhaps Latour does not really want to be so strict; after all, he is allegedly trying to free reality from the self-delusory grip of science, reminding us that alternative theories might, later on, turn out to be right, and reminding science that its grip on reality depends on the constant and laborious practice of scientists. This does not compromise the past existence of microorganisms any more than doubts or confirmations of the accusations of cannibalism, attributed to the Callinago Caribs, deny their existence or their history. So, rather than becoming fixated on what Latour demonstrates to be the highly politicized record of the history of mainstream science, we should pay attention to alternative sources. Although they are likely to be no less politicized, they bring additional perspectives through which we can address the reality and history of microorganisms. The history of old and 'savage' ideas should be examined more carefully, along with the micro-paleontology founded by Ehrenberg,[85] and molecular biological techniques.

Edith Turner,[86] while propounding the validity of and need for experiential anthropological approaches, and acknowledging the realities experienced by 'natives' in regard to spirits, points out that a symptom of Western anthropologists' intellectual imperialism has been to treat these experiences as symbols or metaphors. She then shows how explaining the natives' system in such 'positivistic' terms is hardly different from the methods employed and results achieved by religious missionaries in trying to eradicate native belief systems. Although I agree with her on the similarity between the methods and outcomes of anthropologists and religious missionaries, I would not want to call the former positivists—not only because I doubt the likes of E. Evans-Pritchard or Mary Douglas would accept such a tag, but precisely because when Evans-Pritchard states

> The physiological condition which is said to be the seat of witchcraft, and which I believe to be nothing more than food passing through the small intestine, is an objective condition, but the qualities they attribute to it and the rest of their beliefs about it are mystical.[87]

There is in this treatment of the subject no effort to address the reality of the phenomenon with which he is dealing. His beliefs are by no means positive proofs of what is going on in the small intestine and in the eyes of the Azande, still less of the qualities they attribute to it. Inability to believe in bacteria or their qualities, many of which are awesome, would not make them less real, even if the researcher's lack of apparatus precluded their observation, or if he found microbiologists' descriptions of their qualities hard to believe. The problem, however, is that the burden of proof lies on Evans-Pritchard, the scientist, rather than on the Azande. But his inability not only to believe but to imagine what they were referring to precludes the possibility of understanding their reality in one way or another, and thereby of testing hypotheses about it, thus foreclosing comprehension of the phenomenon and any approach to their reality.

On the other hand, we must recall how frequently tags such as mystical, metaphysical, and supernatural, as much as those like spirits, witchcraft, or demons, are black boxes which resulted from the interactions with Western religious missionaries and their conflicts of power with local authorities, their practices, and their knowledge. This relation generally precedes anthropological inquiry. Pané,[88] for instance, through his dealings starts classifying *buhities* as medical doctors, but soon is turning them into idolatrous quacks. The specialists he is describing are not distant from what contemporary anthropology identifies as shamans, neither perhaps from microbiologists.

Tagging a phenomenon as mystical, metaphysical, or supernatural, or blindly accepting tags such as spirit or witchcraft, frequently obscures the subject. We may regard with suspicion alleged phenomena such as the influence of the sun, the moon, and the planets on contagious diseases, or worse still the evil eye, which were regarded by early empiricist like Roger Bacon as natural phenomena akin to the tides.[89] Nonetheless, modern understanding of the tides and seasonality, or the role of sunlight in photosynthesis and vitamin D metabolism, is a persistent reminder of scientific astrological thought. Moreover, many phenomena, which in the past would have been considered supernatural, now lie within the fences of science. Think about Newton's thoughts on radio waves, or on the notion of energy before Faraday, or on quantum physics of Einstein. Casting a mystical fog is only with difficulty equated with the scientific method or with a positivistic approach. If, on the other hand, we want to challenge the reality 'imposed' by objective science, we must bear in mind that other cultures have already made great contributions to building and multiplying these sciences, and that they neither can nor should be stripped of one another, or reduced to any single one of them.

The homologation of shamanic and microbiological approaches has been hindered by the topic-oriented research in microbiology, by the prejudice that until the development of the microscope humans were unaware of the existence of microbes, and by the idea that Germ theory was a development of the nineteenth century.[90] Still more radically, it has been claimed[91] that microbes did not exist before Pasteur—a view which, although probably accurate for his strains of *Lactobacillus* spp., cannot be extended to every other microorganism. As the **Third Part** will show, a contagion theory which foreshadows the insights of germ theory was proposed shortly after contact with America.[92] The description of microscopic entities since the seventeenth century[93] started corroborating this theory, but our senses actually provide numerous channels of phenomenological access to these worlds. Moreover, our relations with microbes are complex and not restricted to dealing with filth, disease, or contagion. Numerous productive processes such as the fermentation of foods[94] and inebriating drinks and baits, or the retting of fibres to produce ropes[95] actively cultivate and take advantage of microbes.

Incidentally, the domestication of all modern beer yeast strains (*Saccharomyces cerivisiae*) can be traced down to ca. 1570, thousands of years after Europeans started fermenting beverages and hundreds of years before Pasteur, though soon after the Encounter. Ever since that time beer brewers have recycled yeasts, which, by continuously reproducing asexually, lost their ability to reproduce sexually. By contrast, wine producers, who depend on a seasonal crop and cannot recycle yeast, continue to rely on semi-wild strains.[96] As can be seen in Breton, long before Pasteur, the Callinago and their shamans related fermentation with the activity of *acansáncou*. Moreover, as will be seen in the **Third Part**, soon after Contact, Europeans started developing a theory of contagion, involving germ-like entities which reproduce and proliferate, attacking the body and causing disease.

Notes

1. Part of the material in this chapter was accepted for publication in the Journal *Anthropology of Consciousness* Decision on Manuscript ID AoC-07-15-OA-0216, and is due for publication in 2018.
2. Kohn 2013; Taussig 1987, 1993.
3. Castaneda 1968.
4. Langdon 2007; Caicedo 2013.
5. Uribe 2002.
6. Alcaraz López and Tascón 2003; Casas 2003; Morales 1990; Losonczy 2006; Kopenawa and Albert 2013.
7. Fortis 2010; Chapin 1976; Lévi-Strauss 1958, 217; Reichel-Dolmatoff 1996; Viveiros de Castro 2004, 2011.
8. Kopenawa and Albert 2013.
9. Viveiros de Castro 2007; Albert and Gomez 1997, 83.
10. Kopenawa and Albert 2013, 55.
11. Kopenawa and Albert 2013, 57.
12. Kopenawa and Albert 2013, 61.
13. In a similar fashion, *nelekana* designates both the Gunadule shamans (who are also characterized by their permeability to these beings) and a

variety of *purpa*, which are their allied shamanic beings or protectors, enlisted to counter the actions of pathogenic *kukukana* and other antagonistic shamanic beings.

14. Kopenawa and Albert 2013, 493.
15. Kopenawa and Albert 2013, 114.
16. Kopenawa and Albert 2013, 73.
17. Kopenawa and Albert 2013, 99.
18. Kopenawa and Albert 2013, 81–83.
19. *Virola elongata.*
20. Kopenawa's account is framed by his particular gender; it is likely that the perspective of a female shaman would render different values to *xapiri*
21. Kopenawa and Albert 2013, 51–53 & 64–66.
22. Kopenawa and Albert 2013, 81–82.
23. There have been several publications popularizing recent findings in microbiology. I find Blaser (2014) and Yong (2016) to be very accessible and insightful.
24. Hamilton and Lenton 1998; Margulis et al. 1993; Haddock et al. 2010; Nealson and Hastings 2006.
25. Ley et al. 2008; Robinson et al. 2010; Blaser 2010.
26. Burrows et al. 2009.
27. Villarreal 2004.
28. Margulis et al. 2006; Wier et al. 2010; Margulis 1970.
29. Poulin 2002.
30. Falkowski et al. 2008; DeLong et al. 2006.
31. Hastings and Nealson 1977; Hastings 1996; Wilson and Hastings 1998; Nealson and Hastings 2006; Cussatlegras and Le Gal 2007; Latz et al. 2008; Haddock et al. 2010.
32. de Kievit and Iglewski 2000; Nealson and Hastings 2006; Velicer 2003; Hastings and Greenberg 1999; Xavier and Bassler 2003.
33. Ben Jacob et al. 2004, 2012.
34. Hamilton and Lenton 1998.
35. Christner et al. 2008; Pöschl et al. 2010.
36. Velicer 2003; Nadell et al. 2009.
37. Gogarten and Townsend 2005; Thomas and Nielsen 2005.
38. Dunny et al. 2008; Nadell et al. 2009. Evolutionary biologists tend to use terms like: sociality, self-interest, and delusion in a heuristic manner, avoiding the discussion of non-human intentionality. However, the

validity of the models depends on the extrapolation of the notion with all its implications.

39. Hastings and Greenberg 1999; de Kievit and Iglewski 2000; Xavier and Bassler 2003; West et al. 2006.
40. Sender et al. 2016.
41. Gill et al. 2006; Turnbaugh et al. 2007; Blaser 2010.
42. Savage 1977; Blaser 2006; Sekirov et al. 2010.
43. Robinson et al. 2010; Charbonneau et al. 2016.
44. Tannock 2005; Charbonneau et al. 2016.
45. Song et al. 2013.
46. David et al. 2013; Lozupone et al. 2012.
47. Feist et al. 2009; Heinken et al. 2013; Thiele et al. 2013; Gill et al. 2006; Turnbaugh et al. 2007; Blaser 2010.
48. Haley et al. 2012.
49. Guerrero et al. 1986; Velicer and Mendes-Soares 2009.
50. Baumler and Sperandio 2016.
51. Sekirov et al. 2010; Poulin 2002; David et al. 2015; Lyte and Cryan 2014.
52. Hernández 2006; de Oliveira-Neto et al. 2000.
53. Pedersen 2001; Willerslev 2004, 2007; Viveiros de Castro 2007; Humphrey 2007; Pedersen 2007; Stépanoff 2009; Isacsson 1993; Århem 1993; Reichel-Dolmatoff 1975; Vasco Uribe 1985; Viveiros de Castro 2002, Fausto 2007.
54. Tannock 2005, 2007.
55. Kuby 1997.
56. Maier and Watkins 1998.
57. Henderson et al. 1996.
58. Lyte 2013.
59. Dinan et al. 2015; Montiel-Castro et al. 2013.
60. Perry et al. 2016.
61. Hastings and Greenberg 1999; de Kievit and Iglewski 2000.
62. Ingram 2011; Paxson 2008.
63. Xu et al. 2007; James et al. 2004.
64. Santos Granero 2009; Århem 1998; Fausto 2007; Fortis 2010.
65. Powell 2012 [1895], 106; Ravetz and Sardar 1997.
66. Latour 1999, 145–173; Douglas 1984.
67. Douglas 1984.
68. Douglas 1984.

69. Douglas 1984, 36.
70. Douglas 1984, 42.
71. Douglas 1984, 32.
72. Douglas 1984, 9.
73. Apffel-Marglin 1987.
74. Latour 1999, 147.
75. Latour 1999, 146.
76. Latour 1999, 153.
77. Latour 1999, 161.
78. Latour 1999, 154 my italics.
79. Pouchet 1859, 379.
80. Pouchet 1865, 18, 19 French original my translation.
81. Pritchard 1842.
82. Pritchard 1842.
83. 1800 [co1676–1723].
84. Fracastoro 1930 [1546].
85. Pritchard 1842.
86. Turner 2003.
87. Evans-Pritchard 1976, 63.
88. Pané 1988 [1498].
89. Bridges 1964, 142–143.
90. Douglas 1984.
91. Latour 1999.
92. Hudson and Morton 1996; Fracastoro 1863 [1535].
93. van Leeuwenhoek and Hoole(Tr) 1800 [co1676–1723].
94. Paxson 2008.
95. Magnus 2006.
96. Gallone et al. 2016.

Bibliography

Albert, B., and G.G. Gomez. 1997. *Yanomami health: An ethnolinguistic manual*. Belem: PR/McT/CNPq.

Alcaraz López, Gloria Margarita, and Leopoldo Yagari Tascón. 2003. The conceptualization of shamanic healing amongst the Colombian Embera: A process of sociocultural and physiological communication. *Investigación Educación y Enfermeria* 21 (2): 60–78.

Apffel-Marglin, Frédérique. 1987. *Smallpox in two systems of knowledge*. Helsinki: World Institute for Development Economics Research.

Århem, Kaj. 1993. Ecosofia makuna. In *La selva humanizada: ecología alternativa en el trópico húmedo colombiano*, ed. François Correa, 109–126. Bogotá: ICAN/FEN/CEREC.

———. 1998. Powers of place: Landscape, territory and local belonging in Northwest Amazonia. In *Locality and belonging*, ed. Nadia Lovell, 75–98. Oxford: Routledge.

Baumler, Andreas J., and Vanessa Sperandio. 2016. Interactions between the microbiota and pathogenic bacteria in the gut. *Nature* 535 (7610): 85–93. https://doi.org/10.1038/nature18849.

Ben Jacob, Eshel, Israela Becker, Yoash Shapira, and Herbert Levine. 2004. Bacterial linguistic communication and social intelligence. *Trends in Microbiology* 12 (8): 366–372. https://doi.org/10.1016/j.tim.2004.06.006.

Ben Jacob, Eshel, Donald S. Coffey, and Herbert Levine. 2012. Bacterial survival strategies suggest rethinking cancer cooperativity. *Trends in Microbiology* 20 (9): 403–410. https://doi.org/10.1016/j.tim.2012.06.001.

Blaser, M.J. 2006. Who are we? Indigenous microbes and the ecology of human diseases. *EMBO Reports* 7 (10): 956.

Blaser, Martin J. 2010. Harnessing the power of the human microbiome. *Proceedings of the National Academy of Sciences* 107 (14): 6125–6126. https://doi.org/10.1073/pnas.1002112107.

———. 2014. *Missing microbes: How the overuse of antibiotics is fueling our modern plagues*. London: Macmillan.

Bridges, J.H. 1964. *The 'opus majus' of Roger Bacon*, vol. I. Frankfurt: Minerva-Verlag. Original edition, 1266.

Burrows, S.M., W. Elbert, M.G. Lawrence, and U. Pöschl. 2009. Bacteria in the global atmosphere – Part 1: Review and synthesis of literature data for different ecosystems. *Atmospheric Chemistry and Physics* 9 (23): 9263–9280. https://doi.org/10.5194/acp-9-9263-2009.

Caicedo, Alhena. 2013. *La alteridad radical que cura. Los nuevos lugares del chamanismo en Colombia*. Tesis de doctorado. École des Hautes Études en Sciences Sociales, París.[Links].

Casas, Eudoro. 2003. *Practices related to the health-disease process among the Tulekuna: The case of malaria*. Magister en Salud Colectiva MS Colective Health, Facultad de Enfermeria, Universidad de Antioquía.

Castaneda, Carlos. 1968. *The teachings of Don Juan: A Yaqui way of knowledge*. New York: Simon and Schuster.

Chapin, M. 1976. Muu ikala: Kuna birth ceremony. In *Ritual and symbol in Central America*, ed. Philip Young and James Howe, 57–66. Eugene Oregon: University of Oregon.

Charbonneau, Mark R., Laura V. Blanton, Daniel B. DiGiulio, David A. Relman, Carlito B. Lebrilla, David A. Mills, and Jeffrey I. Gordon. 2016. A microbial perspective of human developmental biology. *Nature* 535 (7610): 48–55. https://doi.org/10.1038/nature18845.

Christner, Brent C., Cindy E. Morris, Christine M. Foreman, Rongman Cai, and David C. Sands. 2008. Ubiquity of biological ice nucleators in snowfall. *Science* 319 (5867): 1214–1214. https://doi.org/10.1126/science.1149757.

Cussatlegras, Anne-Sophie, and Patrice Le Gal. 2007. Variability in the bioluminescence response of the dinoflagellate *Pyrocystis lunula*. *Journal of Experimental Marine Biology and Ecology* 343 (1): 74–81.

David, Lawrence A., Corinne F. Maurice, Rachel N. Carmody, David B. Gootenberg, Julie E. Button, Benjamin E. Wolfe, Alisha V. Ling, A. Sloan Devlin, Yug Varma, and Michael A. Fischbach. 2013. Diet rapidly and reproducibly alters the human gut microbiome. *Nature* 505(7484):559–63.

David, G. Biron, Bonhomme Ludovic, Coulon Marianne, and Øverli Øyvind. 2015. Microbiomes, plausible players or not in alteration of host behavior. *Frontiers in Microbiology* 5. doi: https://doi.org/10.3389/fmicb.2014.00775.

de Kievit, Teresa R., and Barbara H. Iglewski. 2000. Bacterial quorum sensing in pathogenic relationships. *Infection and Immunity* 68 (9): 4839–4849. https://doi.org/10.1128/iai.68.9.4839-4849.2000.

de Oliveira-Neto, Manoel, Marise S. Mattos, Mauricio A. Perez, Alda Da-Cruz, Octavio Fernandes, João Moreira, Sylvio Gonçalves-Costa, Lúcia R. Brahin, Claudio R. Menezes, and Claude Pirmez. 2000. American Tegumentary Leishmaniasis (ATL) in Rio de Janeiro state, Brazil: Main clinical and epidemiologic characteristics. *International Journal of Dermatology* 39 (7): 506–514.

DeLong, Edward F., Christina M. Preston, Tracy Mincer, Virginia Rich, Steven J. Hallam, Niels-Ulrik Frigaard, Asuncion Martinez, Matthew B. Sullivan, Robert Edwards, and Beltran Rodriguez Brito. 2006. Community genomics among stratified microbial assemblages in the ocean's interior. *Science* 311 (5760): 496–503.

Dinan, Timothy G., Roman M. Stilling, Catherine Stanton, and John F. Cryan. 2015. Collective unconscious: How gut microbes shape human behavior. *Journal of Psychiatric Research* 63: 1–9. https://doi.org/10.1016/j.jpsychires.2015.02.021.

Douglas, Mary. 1984. *Purity & danger : An analysis of the concepts of pollution & taboo*. London: Routledge.
Dunny, G.M., T.J. Brickman, and M. Dworkin. 2008. Multicellular behavior in bacteria: Communication, cooperation, competition and cheating. *Bioessays* 30 (4): 296–298.
Evans-Pritchard, E.E. 1976. *Witchcraft, oracles and magic among the Azande*. Oxford: Clarendon Press.
Falkowski, Paul G., Tom Fenchel, and Edward F. Delong. 2008. The microbial engines that drive earth's biogeochemical cycles. *Science* 320 (5879): 1034–1039.
Fausto, Carlos. 2007. Feasting on people: Eating animals and humans in Amazonia. *Current Anthropology* 48 (4): 497–530. https://doi.org/10.1086/518298.
Feist, Adam M., Markus J. Herrgård, Ines Thiele, Jennie L. Reed, and Bernhard Ø. Palsson. 2009. Reconstruction of biochemical networks in microorganisms. *Nature Reviews Microbiology* 7 (2): 129–143.
Fortis, Paolo. 2010. The birth of design: A Kuna theory of body and personhood. *Journal of the Royal Anthropological Institute* 16 (3): 480–495.
Fracastoro, Hieronimus. 1863 [1535]. *La Syphilis: poema latino/de Gerónimo Fracastor; traducido al castellano e ilustrado con notas por Luis María Ramírez y de las Casas-Leza*. Madrid: Imprenta de de José M. Ducazal.
Fracastoro, Girolamo. 1930 [1546]. *Hieronymi Fracastorii De contagione et contagiosis morbis et eorum curatione: libri III*. Trans. Haven Emerson, *History of medicine series*. New York: Putnam.
Gallone, Brigida, Jan Steensels, Troels Prahl, Leah Soriaga, Veerle Saels, Beatriz Herrera-Malaver, Adriaan Merlevede, Miguel Roncoroni, Karin Voordeckers, and Loren Miraglia. 2016. Domestication and divergence of saccharomyces cerevisiae beer yeasts. *Cell* 166 (6): 1397–1410.e16.
Gill, S.R., M. Pop, R.T. DeBoy, P.B. Eckburg, P.J. Turnbaugh, B.S. Samuel, J.I. Gordon, D.A. Relman, C.M. Fraser-Liggett, and K.E. Nelson. 2006. Metagenomic analysis of the human distal gut microbiome. *Science* 312 (5778): 1355–1359.
Gogarten, J. Peter, and Jeffrey P. Townsend. 2005. Horizontal gene transfer, genome innovation and evolution. *Nature Reviews Microbiology* 3 (9): 679.
Guerrero, Ricardo, Carlos Pedrós-Alió, Isabel Esteve, Jordi Mas, David Chase, and Lynn Margulis. 1986. Predatory prokaryotes: Predation and primary consumption evolved in bacteria. *Proceedings of the National Academy of Sciences of the United States of America* 83 (7): 2138–2142.

Haddock, Steven H.D., Mark A. Moline, and James F. Case. 2010. Bioluminescence in the sea. *Annual Review of Marine Science* 2 (1): 443–493. https://doi.org/10.1146/annurev-marine-120308-081028.

Haley, Bradd J., Arlene Chen, Christopher J. Grim, Philip Clark, Celia Municio Diaz, Elisa Taviani, Nur A. Hasan, Elizabeth Sancomb, Wessam Mahmoud Elnemr, and Muhammad A. Islam. 2012. Vibrio cholerae in a historically cholera-free country. *Environmental Microbiology Reports* 4 (4): 381–389.

Hamilton, W.D., and T.M. Lenton. 1998. Spora and Gaia: How microbes fly with their clouds. *Ethology Ecology & Evolution* 10 (1): 1–16.

Hastings, J. Woodland. 1996. Chemistries and colors of bioluminescent reactions: A review. *Gene* 173 (1): 5–11.

Hastings, J.W., and E.P. Greenberg. 1999. Quorum sensing: The explanation of a curious phenomenon reveals a common characteristic of bacteria. *Journal of Bacteriology* 181 (9): 2667.

Hastings, J.W., and K.H. Nealson. 1977. Bacterial bioluminescence. *Annual Review of Microbiology* 31 (1): 549–595. https://doi.org/10.1146/annurev.mi.31.100177.003001.

Heinken, Almut, Sahoo Swagatika, R.M. Fleming, and Ines Thiele. 2013. Systems-level characterization of a host-microbe metabolic symbiosis in the mammalian gut. *Gut Microbes* 4 (1): 28–40.

Henderson, Brian, Stephen Poole, and Michael Wilson. 1996. Microbial/host interactions in health and disease: Who controls the cytokine network? *Immunopharmacology* 35 (1): 1–21.

Hernández, C.A. 2006. Historia natural de la leishmaniasis cutánea y mucocutánea. *Biomédica* 26 (1): 10–12.

Hudson, Margaret M., and Robert S. Morton. 1996. Fracastoro and syphilis: 500 years on. *The Lancet* 348 (9040): 1495–1496.

Humphrey, Caroline. 2007. Inside and outside the mirror: Mongolian Shamans' mirrors as instruments of perspectivism. *Inner Asia* 9: 173–197.

Ingram, Mrill. 2011. Fermentation, rot, and other human-microbial performances. In *Knowing nature: Conversations at the intersection of political ecology and science studies*, ed. M.J. Goldman, P. Nadasdy, and M.D. Turner, 99–112. Chicago: University of Chicago Press.

Isacsson, Sven Erik. 1993. *Transformations of eternity*. Göteborg: Department of Social Anthropology, University of Göteborg.

James, A.G., J. Casey, D. Hyliands, and G. Mycock. 2004. Fatty acid metabolism by cutaneous bacteria and its role in axillary malodour. *World Journal of Microbiology and Biotechnology* 20 (8): 787–793.

Jenner, Edward. 1801. *An inquiry into the causes of the variola vaccine, a disease discovered in some of the western counties of England*. London: D.N. Shuby.

Kohn, Eduardo. 2013. *How forests think: Toward an anthropology beyond the human*. Berkeley: University of California Press.

Kopenawa, Davi, and Bruce Albert. 2013. *The falling sky: Words of a Yanomami Shaman*. Trans. Nicholas Elliott and Alison Dundy. Cambridge, MA: Harvard University Press.

Kuby, Janis. 1997. *Immunology*. New York: WH Freeman and Company.

Langdon, Esther Jean. 2007. Shamans and Shamanisms: Refections on anthropological dilemmas of modernity. *Virtual Brazilian Anthropology* 4 (2): 27–48.

Latour, Bruno. 1999. *Pandora's hope: Essays on the reality of science studies*. Cambridge, MA: Harvard University Press.

Latz, Michael I., Michelle Bovard, Virginia VanDelinder, Enrico Segre, Jim Rohr, and Alex Groisman. 2008. Bioluminescent response of individual dinoflagellate cells to hydrodynamic stress measured with millisecond resolution in a microfluidic device. *Journal of Experimental Biology* 211 (17): 2865–2875.

Lévi-Strauss, Claude. 1958. *Structural anthropology*. Paris: Plon.

Ley, R.E., C.A. Lozupone, M. Hamady, R. Knight, and J.I. Gordon. 2008. Worlds within worlds: Evolution of the vertebrate gut microbiota. *Nature Reviews Microbiology* 6 (10): 776.

Losonczy, Anne Marie. 2006. *The interetnic weft*. Bogotá: ICAHN, IFEA.

Lozupone, Catherine A., Jesse I. Stombaugh, Jeffrey I. Gordon, Janet K. Jansson, and Rob Knight. 2012. Diversity, stability and resilience of the human gut microbiota. *Nature* 489 (7415): 220–230.

Lyte, Mark. 2013. Microbial endocrinology in the microbiome-gut-brain axis: How bacterial production and utilization of neurochemicals influence behavior. *PLoS Pathogens* 9 (11): e1003726. https://doi.org/10.1371/journal.ppat.1003726.

Lyte, M., and J.F. Cryan. 2014. *Microbial endocrinology: The microbiota-gut-brain axis in health and disease*. New York: Springer.

Magnus, Ole. 2006. Reconstruction of the rope for the copy of Skuldelev 2: Rope in the Viking Period. In *Connected by the sea*, ed. L.K. Blue, F.M. Hocker, and A. Englert. Oxford: Oxbow Books.

Maier, Steven F., and Linda R. Watkins. 1998. Cytokines for psychologists: Implications of bidirectional immune-to-brain communication for understanding behavior, mood, and cognition. *Psychological Review* 105 (1): 83–107.

Margulis, Lynn. 1970. *Origin of eukaryotic cells*. New Haven: Yale University Press.
Margulis, L., J.B. Ashen, M. Soló, and R. Guerrero. 1993. Composite, large spirochetes from microbial mats: Spirochete structure review. *Proceedings of the National Academy of Sciences of the United States of America* 90 (15): 6966–6970.
Margulis, Lynn, Michael Chapman, Ricardo Guerrero, and John Hall. 2006. The last eukaryotic common ancestor (LECA): Acquisition of cytoskeletal motility from aerotolerant spirochetes in the proterozoic eon. *Proceedings of the National Academy of Sciences* 103 (35): 13080–13085. https://doi.org/10.1073/pnas.0604985103.
Montiel-Castro, Augusto J., Rina M. González-Cervantes, Gabriela Bravo-Ruiseco, and Gustavo Pacheco-López. 2013. The microbiota-gut-brain axis: Neurobehavioral correlates, health and sociality. *Frontiers in Integrative Neuroscience* 7. https://doi.org/10.3389/fnint.2013.00070.
Morales, Jorge. 1990. Fauna: Work and disease among the Kuna. In *The humanized forest: Alternative ecology in the humid Colombian tropic*, ed. François Correa, 167–187. Bogotá: ICAN/FEN/CEREC.
Nadell, C.D., J.B. Xavier, and K.R. Foster. 2009. The sociobiology of biofilms. *FEMS Microbiology Reviews* 33 (1): 206–224.
Nealson, K.H., and J. Hastings. 2006. Quorum sensing on a global scale: Massive numbers of bioluminescent bacteria make milky seas. *Applied and Environmental Microbiology* 72 (4): 2295.
Pané, Ramón Fray. 1988 [1498]. Relación de Fray Ramón acerca de las antigüedades de los indios, las cuales, con diligencia, como hombre que sabe la lengua de ellos, las ha recogido por mandato del Almirante. In *Relación acerca de las antigüedades de los indios: El primer tratado escrito en América*, ed. Jose Juan Arrom. Bógota: Siglo XXI Ediciones.
Paxson, Heather. 2008. Post-pasteurian cultures: The microbiopolitics of raw-milk cheese in the United States. *Cultural Anthropology* 23 (1): 15–47.
Pedersen, Morten A. 2001. Totemism, animism and North Asian indigenous ontologies. *The Journal of the Royal Anthropological Institute* 7 (3): 411–427.
———. 2007. Multiplicity without myth: Theorizing Darhad perspectivism. *Inner Asia* 9 (2): 311–328.
Perry, Rachel J., Peng Liang, Natasha A. Barry, Gary W. Cline, Dongyan Zhang, Rebecca L. Cardone, Kitt Falk Petersen, Richard G. Kibbey, Andrew

L. Goodman, and Gerald I. Shulman. 2016. Acetate mediates a microbiome–brain–β-cell axis to promote metabolic syndrome. *Nature* 534 (7606): 213–217. https://doi.org/10.1038/nature18309. http://www.nature.com/nature/journal/v534/n7606/abs/nature18309.html#supplementary-information.

Pöschl, U., S.T. Martin, B. Sinha, Q. Chen, S.S. Gunthe, J.A. Huffman, S. Borrmann, D.K. Farmer, R.M. Garland, G. Helas, J.L. Jimenez, S.M. King, A. Manzi, E. Mikhailov, T. Pauliquevis, M.D. Petters, A.J. Prenni, P. Roldin, D. Rose, J. Schneider, H. Su, S.R. Zorn, P. Artaxo, and M.O. Andreae. 2010. Rainforest aerosols as biogenic nuclei of clouds and precipitation in the Amazon. *Science* 329 (5998): 1513–1516. https://doi.org/10.1126/science.1191056.

Pouchet, F.A. 1859. *Hétérogénie ou traité de la génération spontanée, basé sur de nouvelles oxpériences: Avec 3 planches gravées*. JB Baillière.

Pouchet, F. A. 1865. *L'univers: les infiniment grands et les infiniment petits*. Librairie de L. Hachette et Cie.

Poulin, Robert. 2002. Parasite manipulation of host behaviour. In *Behavioural ecology of parasites*, ed. E.E. Lewis, J.F. Campbell, and M.V.K. Sukhdeo. Cambridge, MA: CABI Publishing.

Powell, J.W. 2012 [1895]. *The exploration of the Colorado River and its canyons*. New York: Dover Publications.

Pritchard, A. 1842. *A history of infusoria, living and fossil, arranged according to 'die infusionsthierchen' of CG Ehrenberg*. London: Whittaker & CO.

Ravetz, Jerome R., and Ziauddin Sardar. 1997. Rethinking science. *Futures* 29 (6): 467–470.

Reichel-Dolmatoff, G. 1975. *The shaman and the jaguar. A study of narcotic drugs among the Indians of Colombia*. Philadelphia: Temple University Press.

———. 1996. *The forest within: The world-view of the tukano amazonian indians*. San Francisco: Council Oaks Distribution.

Robinson, Courtney J., J.M. Brendan, Bohannan, and Vincent B. Young. 2010. From structure to function: The ecology of host-associated microbial communities. *Microbiology and Molecular Biology Reviews* 74 (3): 453–476. https://doi.org/10.1128/mmbr.00014-10.

Santos-Granero, F. 2009. Hybrid bodyscapes: A visual history of Yanesha patterns of cultural change. *Current Anthropology* 50 (4): 477–512.

Savage, D.C. 1977. Microbial ecology of the gastrointestinal tract. *Annual Reviews in Microbiology* 31 (1): 107–133.

Sekirov, Inna, Shannon L. Russell, L. Caetano, M. Antunes, and B. Brett Finlay. 2010. Gut microbiota in health and disease. *Physiological Reviews* 90 (3): 859–904. https://doi.org/10.1152/physrev.00045.2009.

Sender, Ron, Shai Fuchs, and Ron Milo. 2016. Revised estimates for the number of human and bacteria cells in the body. *bioRxiv*. https://doi.org/10.1101/036103.

Song, Se Jin, Christian Lauber, Elizabeth K Costello, Catherine A Lozupone, Gregory Humphrey, Donna Berg-Lyons, J Gregory Caporaso, Dan Knights, Jose C Clemente, and Sara Nakielny. 2013. Cohabiting family members share microbiota with one another and with their dogs. *Elife* 2. https://doi.org/10.7554/eLife.00458.001.

Stépanoff, C. 2009. Devouring perspectives: On cannibal shamans in Siberia. *Inner Asia (Cambridge)* 11: 283–307.

Tannock, Gerald W. 2005. Commentary: Remembrance of microbes past. *International Journal of Epidemiology* 34 (1): 13–15. https://doi.org/10.1093/ije/dyh380.

———. 2007. What immunologists should know about bacterial communities of the human bowel. *Seminars in Immunology* 19 (2): 94–105.

Taussig, M.T. 1987. *Shamanism, colonialism, and the wild man: A study in terror and healing*. Chicago: University Chicago Press.

———. 1993. *Mimesis and alterity: A particular history of the senses*. London: Routledge.

Thiele, Ines, Neil Swainston, Ronan M.T. Fleming, Andreas Hoppe, Swagatika Sahoo, Maike K. Aurich, Hulda Haraldsdottir, Monica L. Mo, Ottar Rolfsson, and Miranda D. Stobbe. 2013. A community-driven global reconstruction of human metabolism. *Nature Biotechnology* 31 (5): 419–425.

Thomas, Christopher M., and Kaare M. Nielsen. 2005. Mechanisms of, and barriers to, horizontal gene transfer between bacteria. *Nature Reviews Microbiology* 3 (9): 711.

Turnbaugh, P.J., R.E. Ley, M. Hamady, C.M. Fraser-Liggett, R. Knight, and J.I. Gordon. 2007. The human microbiome project. *Nature* 449 (7164): 804–810.

Turner, Edith. 2003. The reality of spirits. In *Shamanism: A reader*, ed. G. Harvey. London: Routledge.

Uribe, Carlos Alberto. 2002. El yajé como sistema emergente: discusiones y controversias. *Documentos del Ceso* 33.

van Leeuwenhoek, Antoni, and Samuel Hoole(Tr). 1800 [co1676–1723]. *The select works of Antony van Leeuwenhoek, containing his microscopical discoveries in many of the works of nature*. London: Sidney and Nicol.

Vasco Uribe, L.G. 1985. *Jaibanás: los verdaderos hombres*. Bogotá: Biblioteca Banco Popular.

Velicer, Gregory J. 2003. Social strife in the microbial world. *Trends in Microbiology* 11 (7): 330–337. https://doi.org/10.1016/S0966-842X(03)00152-5.

Velicer, Gregory J., and Helena Mendes-Soares. 2009. Bacterial predators. *Current Biology* 19 (2): R55–R56.

Villarreal, Luis P. 2004. Can viruses make us human? *Proceedings of the American Philosophical Society* 148 (3): 296–323.

Viveiros de Castro, Eduardo. 2002. *Inconstância da alma selvagem*. São Paulo: Cosac Naify.

———. 2004. Exchanging perspectives: The transformation of objects into subjects in Amerindian ontologies. *Common Knowledge* 10 (3): 463–484.

———. 2007. The crystal forest: Notes on the ontology of Amazonian spirits. *Inner Asia* 9: 153–172.

———. 2011. *The inconstancy of the Indian soul: The encounter of catholics and cannibals in 16th-century Brazil*. Trans. Gregory Duff Morton. Chicago: Prickly Paradigm Press.

West, Stuart A., Ashleigh S. Griffin, Andy Gardner, and Stephen P. Diggle. 2006. Social evolution theory for microorganisms. *Nature Review Microbiology* 4 (8): 597–607.

Wier, Andrew M., Luciano Sacchi, Michael F. Dolan, Claudio Bandi, James Macallister, and Lynn Margulis. 2010. Spirochete attachment ultrastructure: Implications for the origin and evolution of cilia. *The Biological Bulletin* 218 (1): 25–35.

Willerslev, Rane. 2004. Not animal, not *not*-animal: Hunting, imitation and empathetic knowledge among the Siberian Yukaghirs. *The Journal of the Royal Anthropological Institute* 10 (3): 629–652.

———. 2007. *Soul hunters: Hunting, animism, and personhood among the Siberian Yukaghirs*. Berkeley: University of California Press.

Wilson, Thérèse, and J. Woodland Hastings. 1998. Bioluminescence. *Annual Review of Cell and Developmental Biology* 14 (1): 197–230. https://doi.org/10.1146/annurev.cellbio.14.1.197.

Xavier, K.B., and B.L. Bassler. 2003. LuxS quorum sensing: More than just a numbers game. *Current Opinion in Microbiology* 6 (2): 191–197.

Xu, Yun, Sarah Dixon, Richard Brereton, Helena Soini, Milos Novotny, Karlheinz Trebesius, Ingrid Bergmaier, Elisabeth Oberzaucher, Karl Grammer, and Dustin Penn. 2007. Comparison of human axillary odour profiles obtained by gas chromatography/mass spectrometry and skin microbial profiles obtained by denaturing gradient gel electrophoresis using multivariate pattern recognition. *Metabolomics* 3 (4): 427–437.

Yong, Ed. 2016. *I contain multitudes: The microbes within us and a grander view of life*. New York: Random House.

Part II

Shamanic Microscopy, Perceiving Cellular "Souls" and Microbial "Spirits"

In the previous chapters, I have suggested that Amerindian shamanism might be better understood through notions of microbiology, which in the West were only developed after the Encounter, rather than with the metaphysical notions introduced by European missionaries. We have seen that our contemporary understanding of microbes matches the descriptions of entities like the zemes, *acansáncou,* and *utupë.* Microbes share similar attributes with these entities; they participate in the same processes in our body and in the environment. To verify this translation, we still have to address shamanic epistemologies, how do they develop and ground their knowledge, and ascertain whether those epistemologies also afford access to the microbial world.

Shamanic epistemologies rely on multisensory experiences, detailed observations of the environment, the behaviour of the different beings that constitute it, and inquiries relating to the whereabouts, activities, and actions of patients, that might have led to disease or mishap.[1] Furthermore, dreams and hallucinations or visions, rather than being dismissed as delusions or symbolic constructs, are recognized as means of perceptual access to physical reality. Lowland South American shamans

---

Some of the material in this Part was accepted for publication in the Journal *Anthropology of Consciousness,* decision on Manuscript ID AoC-07-15-OA-0216, and is due for publication in 2018.

claim to be able to diagnose and treat infectious diseases, as well as to assess the status of wildlife resources and carry environmental interventions through interactions with pathogenic agents perceived in visions and dreams.

It has been suggested that part of what is experienced in shamanic visions are phosphenes or entoptics, phenomena arising from events occurring within the visual system. Psychologists and anthropologists have emphasized the alleged neuronal origin of these phenomena, which presumably reveal the underlying workings of the mind.[2] However, early and recent physiological research expose perplexing features of the retina that afford a form of microscopy through entoptic vision (of events within the eye). Entoptic microscopy allows the perception of the viewers' own retinal structures, blood cells, microscopic particles flowing through retinal capillaries, and of microbes during systemic infections. Entoptic microscopy is optimized under optical and physiological conditions that feature prominently in shamanic practices. Moreover, the depictions of shamanic visions display the characteristic features of entoptic microscopy. Shamanic visions may be a subjective means to engage with microbes, entities recognized by biomedicine as the causal agents of many of the infectious diseases shamans claim to diagnose and treat, which, moreover, are crucial players in the working balance of the ecological communities they claim to manage.

Notes

1. Jokic 2015; Morales 1990, 1994.
2. Reichel-Dolmatoff 1996; Lewis-Williams and Dowson 1990; Lewis-Williams 2004; Lewis-Williams et al. 1988; Reichel-Dolmatoff 1975; Klüver 1926.

Bibliography

Jokic, Zeljko. 2015. *The living ancestors: Shamanism, cosmos and cultural change among the Yanomami of the upper Orinoco*. Oxford: Berghahn.

Klüver, Heinrich. 1926. Mescal visions and eidetic vision. *The American Journal of Psychology* 37 (4): 502–515.

Lewis-Williams, J.D. 2004. Neuropsychology and upper palaeolithic art: Observations on the progress of altered states of consciousness. *Cambridge Archaeological Journal* 14 (01): 107–111. https://doi.org/10.1017/S0959774304000083.

Lewis-Williams, J.D., and Thomas A. Dowson. 1990. On neuropsychology and shamanism in rock art. *Current Anthropology* 31 (1): 77–84.

Lewis-Williams, J.D., T.A. Dowson, Paul G. Bahn, H.G. Bandi, Robert G. Bednarik, John Clegg, Mario Consens, Whitney Davis, Brigitte Delluc, Gilles Delluc, Paul Faulstich, John Halverson, Robert Layton, Colin Martindale, Vil Mirimanov, Christy G. Turner, Joan M. Vastokas, Michael Winkelman, and Wylie Alison. 1988. The signs of all times: Entoptic phenomena in upper palaeolithic art [and comments and reply]. *Current Anthropology* 29 (2): 201–245.

Morales, Jorge. 1990. Fauna. Trabajo y enfermedad entre los cuna. In *La selva humanizada: Ecología alternativa en el trópico húmedo colombiano*, ed. François Correa, 167–187. Bogotá: ICAN/FEN/CEREC.

Morales, Diana Marcela. 1994. Enfermedad, curación y jaibanismo. Concepciones embera sobre las enfermedades más comunes. *Boletin del Instituto Francés de Estudios Andinos* 23 (2): 317–357.

Reichel-Dolmatoff, G. 1975. *The shaman and the jaguar. A study of narcotic drugs among the Indians of Colombia*. Philadelphia: Temple University Press.

———. 1996. *The forest within: The world-view of the tukano amazonian indians*. San Francisco: Council Oaks Distribution.

5

Shamanic Epistemologies

Peoples like the Taino and the Callinago had a clear inclination towards a utilitarian empiricism, acknowledging as real solely what they could see and use. Since the first encounters, their shamans described shamanic beings like zemes and *acansáncou*, with many features in common with our contemporary understanding of microbes. In some circumstances, they inferred their presence or influence from the abnormal behaviours of plants and animals, or from signs of disease, which they correlated with interactions with particular places, animals, plants, rocks, and artefacts. They also claimed to perceive these entities directly through smell, through the pulse, through ecstatic visions, aural experiences, and through dreams. Given the proclivities of the Callinago we can assume that they acknowledged these shared experiences as real because they found them useful.

Various Amerindian, Circumpolar, and Southeast Asian peoples have regarded dreams and visions as powerful means of perceptual access to reality, as realms of experience through which humans and non-humans may establish meaningful interactions with other beings. Frequently, their patients or clients, as well as hunters in the forest, children, and outsiders in a community make similar claims.[1] Animists, especially

C. E. Giraldo Herrera, *Microbes and Other Shamanic Beings*,
https://doi.org/10.1007/978-3-319-71318-2_5

shamans, substantiate part of their knowledge on the experiences of visions and dreams. Although such experiences could ground concerns about the fallibility of the senses, animists value the reality of these experiences and correlate the perceptual capacities of their shamans with a better vision, which, rather than fooling them, affords them a deeper perspective into reality, literally allowing them to see within the body. This chapter examines the role of visions and dreams as sources of knowledge for Amerindian shamans.

In animist ontologies such as those of the Callinago, the body is permeable to a variety of beings, which may affect behaviour and perception. Consequently, dreaming is more than the epiphenomena of physical processes of a semi-inert body or the past experiences of a brain shut in a vat. For animists, dreams and visions are real and potentially intersubjective experiences. Dreamers are persons. Dog dreaming and their response to hallucinogenic substances evidence, for the Runa peoples of Ecuador, the subjectivity of their dogs.[2] Moreover, animists will frequently acknowledge personhood to entities appearing in these states. Furthermore, although most animists consider that dreams require interpretation, human and non-human persons may be accountable for what they do in their own and in others' dreams and visions.[3]

For long Amerindian and mestizo shamans have employed visions to interact with masters of game and other pathogenic agents, to diagnose and treat infectious and mental diseases, and to monitor and intervene on the environment. The sixteenth-century Taino *buhities* claimed to interact with zemes, through visions induced by inhaled tobacco, and substances like *güeyo*. A hundred years later, the Callinago claimed to perceive *acansáncou* through smell, through the pulse, and as visions and aural experiences induced by fasting and strong tobacco. The Callinago *boyáicou* employed animal substances and songs to call these beings, which appeared as dancing and singing images/shadows descending from the sky.

The images of hallucinations and dreams substantiate important aspects of Amerindian ontologies. Yanomami notion of u*tupë* has been translated 'as image, vital principle, true interiority or essence',[4] or the Kuna *purpa*, the image, secret essence, or soul.[5] Strictly speaking, *utupë* and *purpakana* refer to multitudinous entities, with their own personhood and intentionality that inhabit, mirror, and holographically constitute each being of the environment: humans, animals, plants, soils, bodies of water, rocks, and even tools. This multiplicity of inner images/beings can be perceived in visions and dreams; thus, these notions have a strong visual component. In a wider sense, they also denote shadows, scintillations, reflections, photographs, and TV images. Nonetheless, *acansáncou, purpakana* and *utupë* also have aural, tactile, proprioceptive, and odorous expressions.

Like the Callinago *boyáicou*, Yanomami shamans employ animal *utupë*, animal substances, and songs learnt from the *xapiri*, or masters of game, to call them and make them descend. The Yanomami recognize multiple other substances that either attract or make *xapiri* visible, including honey. However, the main substance employed by Yanomami shamans is snuff from the *yakoana* tree (*Virola elongata*). With the aid of *yakoana*, trained shamans can see myriads of them dancing, shining their radiant colours, and singing joyous songs,

> their paths look like spider-webs shining like moonlight…The *xapiripë* dance together on huge mirrors which descend from the sky.[6]

Seeing *utupë*, and related entities, implies the mesmerizing notion of seeing within the body. It is a capacity, facilitated by fasting regimes, substances like yage (brews of *Banisteriopsis* spp. in combination with other plants like *Psychotria* spp., *Diplopterys* spp., *Brugmansia, Datura* spp.), snuffs of *Anadenanthera* spp. or *Virola* spp., abundant doses of honey, strong corn or manioc beer, or tobacco, the intense red light of resin-covered torches like the Tukanoan *turí* or the Guandule *sianala* braziers, and instruments like divining crystals and fans.[7] The visions developed through these means fundament the medical and environmental knowledge of Lowland shamans in the Amazonia, the Pacific Coast of Ecuador, Colombia, and the region of Darien between Colombia and Panama.

Although there are profound differences between the practices and the knowledge articulated by particular shamans, they are also renowned for travelling to remote locations, crossing ethnic and social barriers, to treat patients, acquire medicinal substances, or receive training from other shamans.[8] Similar experiences, ontologies, understandings of disease, and means of diagnoses are found through Lowland South America and beyond. In various contexts, the degree of achievement of shamans is correlated with the level of detail of their observations and their depictions, which is frequently correlated with the amount of psychotropic substances they are able to tolerate.[9]

Amongst the Colombian Ingano, a Quechua-speaking group on the Amazonian foothills of the Andes and neighbouring Amerindian and mestizo communities, visions play an important part of shamanism. Shamanic visions are known as *Muskui ukupsina.* This term is translated into Spanish as pintas or paintings; literally, it means: "looking within" or "looking into the inside world". *Muskui ukupsina* is a capacity, facilitated by strong corn beer and by brews of yage (*Banisteriopsis* spp. plus other ingredients), by instruments like the *waira wawa* (divining crystals) and *waira-sacha* (leaf-fans), and by the intense red light of resin-covered torches. Through these means, *curacas* (healers or shamans) and patients engage with *anima waira* and *kuku* pathogenic masters of game. This interaction is part of the therapeutic process.[10] These visions have been illustrated through the work of world-renowned Ingano artist Carlos Jacanamijoy.

Son of a respected *curaca*, much of Jacanamijoy's work is based on his personal interaction with yage. However, he does not claim to be a shaman. Jacanamijoy struggled to be recognized as an artist independently of his ethnic background while simultaneously attempting to preserve and foster his cultural heritage. The result of this position and contradictory trends is a rich artistic language, which, like 'internal navigation' (see Fig. 5.1), is characterized by a series of ethereal and evanescent motifs superposed over figurative images. The motifs depicted by Jacanamijoy share structural similarities with those of other Amazonian and ayahuasca (yage) artists such as Pablo Amaringo, Juan Carlos Taminchi, Sheroanawe Hakihiiwee, or Jheferson Saldaña Valera.[11] However, unlike most of the latter, Jacanamijoy explores the power of these motifs as highly evocative abstract shapes rather than attempting to resolve them as figurative symbols. Jacanamijoy portrays the worlds of yage beautifully, and, I will argue, with great realism.

Now the question is, if, as animists ask from us, we take seriously that what they see, or perceive by other means, are not just symbolic constructs, but real entities (real like a jaguar is real, which is but not limited to Western biological knowledge about *Panthera onca*), what could they be perceiving? How could they see within? Why is detail important? And,

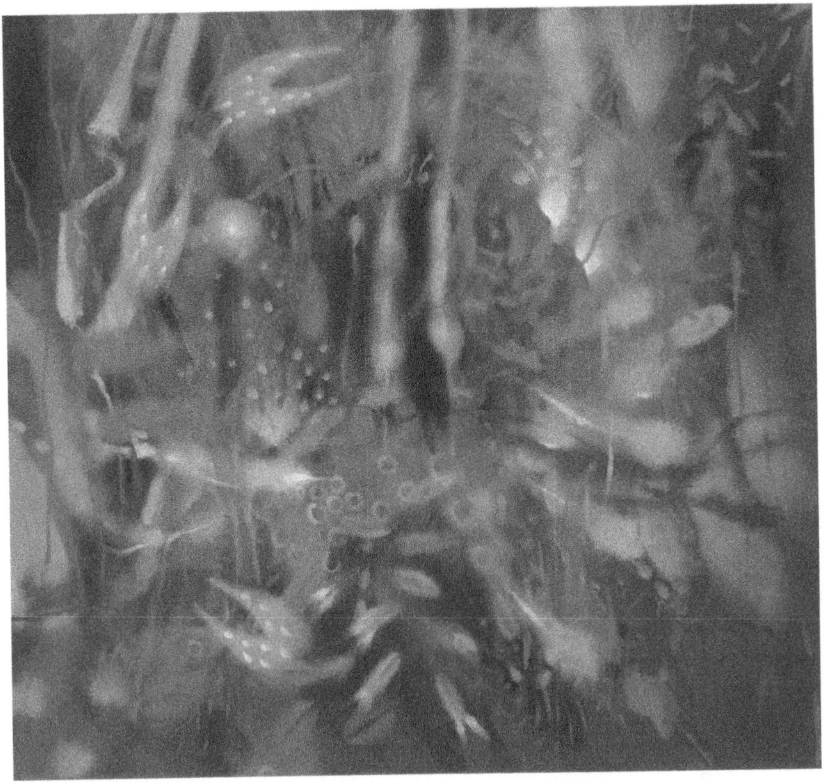

Fig. 5.1 *Navegación interna* (2009), oil on canvas (0.94 × 1.02 M.) by Carlos Jacanamijoy (Reproduced with the kind permission of the author)

how does this relate to disease or to the environment? I will explore some responses through the insights and the blind spots of some naturalistic explanations to shamanic visions, and the potential to build commensurability, drawing further into these explanations.

Notes

1. Chapin 1983, 91 & 101; Viveiros de Castro 2002, 79; Reichel-Dolmatoff 1975, 1996; Willerslev 2004, 2007; Århem 1993, 1996; Delaplace and Empson 2007; Kopenawa and Albert 2013.
2. Kohn 2013, 106, 142–148.

3. Kohn 2007; Bacigalupo 2016; Jokic 2015; Hallowell 2002 [1960]; Niño 2007.
4. Viveiros de Castro 2007; Albert and Gomez 1997, 83.
5. Morales 1990; Chapin 1983; Severi 1981, 1987, 1993; Hayans et al. 1958.
6. Kopenawa and Albert 2013.
7. Reichel-Dolmatoff 1996; Chapin 1983; Kopenawa and Albert 2013.
8. Schindler and Faust 2000; Faust 2001; Reichel-Dolmatoff 1975, 160–166; Taussig 1993; Uribe 2002; Vasco 1985; Losonczy 2006; Gow 1994.
9. MacLean 2001; Reichel-Dolmatoff 1975, 168–181; Kopenawa and Albert 2013; Vasco Uribe 1985.
10. Ronderos Valderrama 2004; Uribe 2002; Schultes 1957; Luna 2011.
11. Arte_Amazonia 2015.

Bibliography

Albert, B., and G.G. Gomez. 1997. *Yanomami health: An ethnolinguistic manual.* Belem: PR/McT/CNPq.

Århem, Kaj. 1993. Makuna ecosophy. In *The humanized forest: Alternative ecology in the humid Colombian tropic*, ed. François Correa, 109–126. Bogotá: ICAN/FEN/CEREC.

———. 1996. The cosmic food web: Human-nature relatedness in the Northwest Amazon. In *Nature and society: Anthropological perspectives*, ed. Philippe Descola and Gisli Pálsson, 185–204. London: Routledge.

Arte_Amazonia. 2015. *Arte Amazonia.* http://www.arte-amazonia.com/amazonia-artists/painting/. Accessed 19 July 2015.

Bacigalupo, Ana Mariella. 2016. *Thunder shaman: Making history with Mapuche spirits in Chile and Patagonia.* Austin: University of Texas Press.

Chapin, N. 1983. *Curing among the San Blas Kuna of Panama.* Doctor in philosophy with a major in cultural anthropology, University of Arizona.

Delaplace, Grégory, and Rebecca Empson. 2007. The little human and the daughter- in-law: Invisibles as seen through the eyes of different kinds of people. *Inner Asia* 9 (2): 197–214.

Faust, Franz Xaver. 2001. Indigenous Cauca. In *Possible territories: History, geography and culture of Cauca*, ed. Guido Barona and Cristóbal Gnecco, 259–274. Popayán: Universidad del Cauca.

Gow, Peter. 1994. River people: Shamanism and history in Western Amazonia. In *Shamanism, history and the state*, ed. Nicholas Thomas and Caroline Humphrey. Ann Arbor: University of Michigan Press.
Hallowell, A.I. 2002 [1960]. Ojibwa ontology, behavior, and world view. In *Readings in Indigenous Religions*, 18–49. London: Continuum.
Hayans, Guillermo, S. Henry Wassén, and Niels M. Holmer. 1958. *Nia-ikala: Magical chant to cure madness: Text in Kuna, annotated by the Indian Guillermo Hayans with Spanish translations and comments*. Gothenburg: Etnografiska museet.
Jokic, Zeljko. 2015. *The living ancestors: Shamanism, cosmos and cultural change among the Yanomami of the upper Orinoco*. Oxford: Berghahn.
Kohn, Eduardo. 2007. How dogs dream: Amazonian natures and the politics of transspecies engagement. *American Ethnologist* 34 (1): 3–24.
———. 2013. *How forests think: Toward an anthropology beyond the human*. Berkeley: University of California Press.
Kopenawa, Davi, and Bruce Albert. 2013. *The falling sky: Words of a Yanomami Shaman*. Trans. Nicholas Elliott and Alison Dundy. Cambridge, MA: Harvard University Press.
Losonczy, Anne Marie. 2006. *The interetnic weft*. Bogotá: ICAHN, IFEA.
Luna, Luis Eduardo. 2011. Indigenous and mestizo use of ayahuasca: An overview. In *The ethnopharmacology of ayahuasca*, ed. Rafael Guimarães dos Santos, 1–21. Kerala: Transworld Research Network.
MacLean, Hope. 2001. Sacred colors and Shamanic vision among the Huichol Indians of Mexico. *Journal of Anthropological Research* 57 (3): 305–323.
Morales, Jorge. 1990. Fauna: Work and disease among the Kuna. In *The humanized forest: Alternative ecology in the humid Colombian tropic*, ed. François Correa, 167–187. Bogotá: ICAN/FEN/CEREC.
Niño, Juan Camilo. 2007. *Ooyoriyasa. Cosmología e interpretación onírica entre los ette del norte de Colombia*. Bogotá: Universidad de los Andes.
Reichel-Dolmatoff, G. 1975. *The shaman and the jaguar. A study of narcotic drugs among the Indians of Colombia*. Philadelphia: Temple University Press.
———. 1996. *The forest within: The world-view of the tukano amazonian indians*. San Francisco: Council Oaks Distribution.
Ronderos Valderrama, Jorge. 2004. El yagé, los curacas (taitas, payés o chamanes) en el Amazonas y el Alto y Bajo Putumayo. Brujería y magia. *Cultura y droga* 11: 91–113.
Schindler, Helmut, and Franz X. Faust. 2000. Interethnic relations amongst healers in the Colombian Southwestern. *Anthropologica* (18–19): 281–294.

Schultes, Richard Evans. 1957. The identity of the malpighiaceous narcotics of South America. *Botanical Museum Leaflets, Harvard University* 18 (1): 1–56.

Severi, Carlo. 1981. The Kuna soul. *La Ricerca Folklorica* 4: 69–75.

———. 1987. The invisible path: Ritual representation of suffering in Kuna traditional thought. *Res Anthropology and Aesthetics* 14: 66–85.

———. 1993. Talking about souls: The pragmatic construction of meaning in Kuna ritual. In *Cognitive aspects of religious symbolism*, ed. Pascal Boyer, 165–172. Cambridge: Cambridge University Press.

Taussig, M.T. 1993. *Mimesis and alterity: A particular history of the senses.* London: Routledge.

Uribe, Carlos Alberto. 2002. *Yage as an emergent system: Discussions and controversies*, vol. 33, *Documentos del Ceso*. Bogota: Ediciones Uniandes.

Vasco, Luis Guillermo. 1985. *Jaibanas: The true men*. Bogota: Biblioteca Banco Popular.

Vasco Uribe, L.G. 1985. *Jaibanás: los verdaderos hombres*. Bogotá: Biblioteca Banco Popular.

Viveiros de Castro, Eduardo. 2002. *The inconstancy of the savage soul.* São Paulo: Cosac Naify.

———. 2007. The crystal forest: Notes on the ontology of Amazonian spirits. *Inner Asia* 9: 153–172.

Willerslev, Rane. 2004. Not animal, not *not*-animal: Hunting, imitation and empathetic knowledge among the Siberian Yukaghirs. *The Journal of the Royal Anthropological Institute* 10 (3): 629–652.

———. 2007. *Soul hunters: Hunting, animism, and personhood among the Siberian Yukaghirs*. Berkeley: University of California Press.

6

Neuropsychological Naturalistic Explanations of Shamanic Visions

Since the earliest encounters, Europeans dismissed the epistemology of shamans and described their experiences as the result of delusions derived from the malfunction of the brain induced by mind-altering substances. Derisively, Pané declares that inebriation conduced *behiques* to fool themselves and others:

> He [The shaman] tells of the vision he had, drunk with the *cohoba* inhaled through the nose, which has gone to his head, and says he has spoken to the zeme, and that there will be victory… or such other thing, which being drunk he remembers. Do judge the state of his brain, if he claims that the house was turned with its foundations upside down, and that men walked with their feet towards the sky.[1]

In a similar fashion, some anthropologists of religion disdained the epistemologies of shamans, and dismissed their practices as forms of delusion and self-delusion.[2] In a subtler manner, various symbolic anthropologists interpreted entities like zemes, *acansáncou,* and *utupë* as sophisticated metaphoric constructs, representing human society, which shamans manipulate to operate over human social relations, psychoanalysing their patients.[3]

C. E. Giraldo Herrera, *Microbes and Other Shamanic Beings*,
https://doi.org/10.1007/978-3-319-71318-2_6

The effectiveness of most medical treatments requires the confidence of the patients. Like other medical specialists,[4] shamans employ subtle means of delusion to acquire power over their patients and treat multiple illnesses through psychoanalytic or psychotherapeutic means.[5] However, the patients' confidence also depends on the effectiveness of the treatments, and therefore on the practitioner's ability to understand and diagnose the diseases. Delusion and psychological manipulation do not explain the phenomena experienced by shamans or their patients, nor do they ascertain the relevant knowledge derived from these phenomena; therefore, these theories fail to address how shamanic healing practices and environmental interventions could be operational.

The interpretations of the anthropology of religion and symbolic anthropology about shamanism have had a complementary, naturalistic explanation, which, like Pané's, have assumed that the substances employed by some Amerindian shamanic traditions produce altered states of the brain.

Several of the substances employed in Amerindian shamanism contain either serotonin (5-hydroxytryptamine) analogues or serotonin blocker inhibitors. Serotonin is a neurotransmitter involved in the regulation of personality, mood, learning, and wakefulness, amongst numerous other neuroendocrine processes. Serotonin analogues emulate the action of serotonin, interacting with its receptors (5-HT), thereby interfering with the regulation of these processes. Serotonin analogs like N, N-Dimethyltryptamine (DMT) are found in the *Psychotria* spp. of yage brews, in bufotenine (found in yopo), mescaline, and psilocybin. On the other, hand serotonin blocker inhibitors interfere with the channels, regulating the availability of serotonin. Type-A monoamine oxidase (MAO-A) normally deaminates (decomposes) serotonin and other neurotransmitters, regulating their availability. However, MAO-A inhibitors (MAO-Ai) halt the deamination, increasing the availability of serotonin and its analogues, potentiating the effects of substances like DMT and bufoteine. MAO-Ai are found in the *Banisteriopsis caapi* of yage and in *Virola* spp. snuffs.[6] Thus, it is understood that yage and *yopo* alter the regulation of multiple neuroendocrine processes and produce various effects that mimic psychotic states, including 'perceptual distortions'[7]: hallucinations, entoptics, and phosphenes.[8]

Hallucinations have been defined as perceptual phenomena produced in the absence of congruent objective stimuli. This definition, however, can be problematic, as many of the phenomena that could be classified as hallucinations are entoptics. These are phenomena resulting from entoptic vision, the subjective observation of events within the eye. This form of vision results from the perception of congruent objective stimuli within the eye, which are not readily apparent to external observers without equipment. Then there are phosphenes, visual phenomena characterized by their brilliance and their visibility in twilight, in near complete darkness, or in the apparent absence of luminous stimuli. Some phosphenes have been identified as entoptics, while others structurally resemble entoptic phenomena. Like various entoptic phenomena, phosphenes are observable in circumstances implying physiological strain, such as fasting, sleeplessness, exhaustion, pain, or meditation. They are evidenced after or through abrupt positional or metabolic changes, and are apparent when there are alterations in systemic or retinal blood pressure, decrease in sugar, or increases in CO_2 concentration. Phosphenes are symptomatic of various ocular and systemic pathological conditions such as retinal or systemic infections, and are also caused by many drugs, including various psychoactive ones, but also a long list of antibiotics.[9] Phosphenes can also be elicited by electromagnetic stimulation of the retina, the optical nerve, or the visual cortex. Although, the causal origin of phosphenes—whether optic or neurogenic—continues to be the source of heated disputes in physiology and neuropsychology, many anthropologists have embraced them as a naturalistic explanation for shamanic experience.

Reichel-Dolmatoff reported phosphenes amongst the effects of *yage* and noted the similarity between the designs drawn from shamanic experiences by the Tukano in the Colombian Vaupes River and the shapes of Knoll's phosphenes—induced by electric currents, of magnitudes resembling those of the neuronal system, applied on the scalp.[10] In a similar fashion, Blackburn noted the similarity of the patterns observed by the shamans of the Chumash of California and phosphenes.[11] Later, in a series of highly influential articles, Lewis-Williams and Dowson explained shamanic experience of the San in South Africa, through the articulation of entoptic phenomena within the neuropsychology of trance experience.[12] Lewis-Williams and Dowson proposed a diagnostic means to identify trance experiences comprising three stages, not necessarily consecutive, which include:

1) Entoptics, which they consider to encompass not only optic phenomena within the eye but also phenomena arising from the neural processes on the retina, the optic nerve, and the visual cortex. However, the authors focus their exploration on neural visual processes, while optic phenomena occurring within the eye were left for future analysis.
2) Geometrical patterns which, they argued, were generated in neural pathways.
3) Iconics which, they assumed, to be cultural constructs articulating perception, generated in the frontal cortex.

After Lewis-Williams and Dowson, others have acknowledged the role of phosphenes or entoptics among different shamanic traditions, like those of the Huichol in the Sierra Madre of Mexico[13] or the Maki in Malakula, Melanesia.[14] In many of these contexts, phosphenes or entoptics are associated with elaborate pictorial expressions. This association has generated a substantial debate in the interpretation of rock and cave art iconography, as it was suggested that the similarity of motifs in different contexts could be derived from the observation of entoptic phenomena. This in turn was considered as suggestive of the involvement of shamanism in various contexts such as Paleocene European rock art[15] and Iron Age European coins.[16]

Some authors[17] have objected that neuropsychological explanations neglect the widely differing circumstances, the diverse substances and paraphernalia employed by shamans from different traditions. Others point out that this interpretation fails to acknowledge the diverse and characteristic patterns perceived by each particular community. As will be shown, these objections do not necessarily refute the theory. Nonetheless, neuropsychological theory of entoptics, as it stands, fails to explain the connection of entoptics with shamanism: what does a shaman, as a healer or an environmental manager, gain from seeing entoptics? To elucidate the potential importance of entoptics for shamans, it is necessary to examine carefully what entoptics are, and to do this, it is necessary to understand how the retina is organized and how it works.

Notes

1. My translation from Pané 1988 [1498], 42.
2. Durkheim 1969 [1915]; Kehoe 2000.
3. Lévi-Strauss 1958, 208; Durkheim 1969 [1915]; Douglas 1984; Evans-Pritchard 1976; Kehoe 2000.
4. Hsu 2000.
5. Taussig 1993; Kehoe 2000; Stépanoff 2009.
6. Schultes et al. 1969.
7. Nichols 2004.
8. Hofmann and Schultes 1993; Riba et al. 2002; Callaway et al. 1999.
9. Cervetto et al. 2007.
10. Reichel-Dolmatoff 1975, 168–181.
11. Blackburn 1977.
12. Lewis-Williams 2004; Lewis-Williams et al. 1988; Dowson 1989.
13. MacLean 2001.
14. Wallis 2002.
15. Lewis-Williams 2004; Lewis-Williams et al. 1988; Hodgson 2000.
16. Creighton 1995.
17. In the discussion to Lewis-Williams et al. 1988.

Bibliography

Blackburn, T. 1977. Biopsychological aspects of Chumash rock art. *Journal of California Anthropology* 4 (1): 88–94.

Callaway, J.C., D.J. McKenna, C.S. Grob, G.S. Brito, L.P. Raymon, R.E. Poland, E.N. Andrade, E.O. Andrade, and D.C. Mash. 1999. Pharmacokinetics of Hoasca alkaloids in healthy humans. *Journal of Ethnopharmacology* 65 (3): 243–256. https://doi.org/10.1016/S0378-8741(98)00168-8.

Cervetto, L., G.C. Demontis, and C. Gargini. 2007. Cellular mechanisms underlying the pharmacological induction of phosphenes. *British Journal of Pharmacology* 150 (4): 383–390.

Creighton, John. 1995. Visions of power: Imagery and symbols in late iron age Britain. *Britannia* 26: 285–301.

Douglas, Mary. 1984. *Purity and danger: An analysis of the concepts of pollution and taboo*. London: Routledge.

Dowson, Thomas A. 1989. Dots and dashes: Cracking the entoptic code in bushman rock paintings. *Goodwin Series* 6: 84–94.

Durkheim, E. 1969 [1915]. *The elementary forms of the religious life*. Trans. Joseph Ward Swain. New York: Free Press.
Evans-Pritchard, E.E. 1976. *Witchcraft, oracles and magic among the Azande*. Oxford: Clarendon Press.
Hodgson, Derek. 2000. Shamanism, phosphenes, and early art: An alternative synthesis. *Current Anthropology* 41 (5): 866–873. https://doi.org/10.1086/317415.
Hofmann, Albert, and Richard Evans Schultes. 1993. *Plantas de los dioses: orígenes del uso de los alucinógenos*. México: FCE.
Hsu, Elisabeth. 2000. Spirit (shen), styles of knowing, and authority in contemporary Chinese medicine. *Culture, Medicine and Psychiatry* 24 (2): 197–229.
Kehoe, A.B. 2000. *Shamans and religion: An anthropological exploration in critical thinking*. Long Grove: Waveland.
Lévi-Strauss, Claude. 1958. *Structural anthropology*. Paris: Plon.
Lewis-Williams, J.D. 2004. Neuropsychology and upper palaeolithic art: Observations on the progress of altered states of consciousness. *Cambridge Archaeological Journal* 14 (01): 107–111. https://doi.org/10.1017/S0959774304000083.
Lewis-Williams, J.D., T.A. Dowson, Paul G. Bahn, H.G. Bandi, Robert G. Bednarik, John Clegg, Mario Consens, Whitney Davis, Brigitte Delluc, Gilles Delluc, Paul Faulstich, John Halverson, Robert Layton, Colin Martindale, Vil Mirimanov, Christy G. Turner, Joan M. Vastokas, Michael Winkelman, and Wylie Alison. 1988. The signs of all times: Entoptic phenomena in upper palaeolithic art [and comments and reply]. *Current Anthropology* 29 (2): 201–245.
MacLean, Hope. 2001. Sacred colors and Shamanic vision among the Huichol Indians of Mexico. *Journal of Anthropological Research* 57 (3): 305–323.
Nichols, David E. 2004. Hallucinogens. *Pharmacology & Therapeutics* 101 (2): 131–181. https://doi.org/10.1016/j.pharmthera.2003.11.002.
Pané, Ramón Fray. 1988 [1498]. Relación de Fray Ramón acerca de las antigüedades de los indios, las cuales, con diligencia, como hombre que sabe la lengua de ellos, las ha recogido por mandato del Almirante. In *Relación acerca de las antigüedades de los indios: El primer tratado escrito en América*, ed. Jose Juan Arrom. Bógota: Siglo XXI Ediciones.
Reichel-Dolmatoff, G. 1975. *The shaman and the jaguar. A study of narcotic drugs among the Indians of Colombia*. Philadelphia: Temple University Press.
Riba, Jordi, Peter Anderer, Adelaida Morte, Gloria Urbano, Francesc Jané, Bernd Saletu, and Manel J. Barbanoj. 2002. Topographic pharmaco-EEG

mapping of the effects of the South American psychoactive beverage ayahuasca in healthy volunteers. *British Journal of Clinical Pharmacology* 53 (6): 613–628. https://doi.org/10.1046/j.1365-2125.2002.01609.x.

Schultes, Richard Evans, Bo Holmstedt, and Jan-Erik Lindgren. 1969. De plantis toxicariis e mundo novo tropicale commentationes III: Phytochemical examination of Spruce's original collection of Banisteriopsis caapi. *Botanical Museum Leaflets, Harvard University* 22 (4): 121–132.

Stépanoff, C. 2009. Devouring perspectives: On cannibal shamans in Siberia. *Inner Asia* 11: 283–307.

Taussig, M.T. 1993. *Mimesis and alterity: A particular history of the senses*. London: Routledge.

Wallis, Robert J. 2002. The bwili or 'flying tricksters' of malakula: A critical discussion of recent debates on rock art, ethnography and shamanisms. *The Journal of the Royal Anthropological Institute* 8 (4): 735–760.

7

The Cavern of the Eye: Seeing Through the Retina

The human eye is a wonderful evolutionary achievement; this is evident, among many other things, in its extreme sensibility to luminosity, as well as in its versatility to work under a wide range of lighting conditions. In a normal human retina, there are about 90 million rods, photoreceptors adapted to low-luminosity conditions, and five million cones, cells adapted to higher luminosity colour vision. Individual rods reach the absolute limit of sensibility to the detection of light, cast by quantum and molecular theories.[1] In our everyday life, our photoreceptors adapt successfully to more than ten orders of magnitude (10^{-4}–10^{6} photons/second). This requires not only being able to count photons, like the photomultipliers with which physicists work in their laboratories, but also being able to catch the right photon and match it to a real source.[2] The versatility of the eye is achieved to a great degree through macrostructures like the pupil acting as a diaphragm filtering light, the crystalline concentrating it, and the vitreous humour conducting it optimally. Of equal or greater importance, though, are the specialization, organization, and interconnection of the different photoreceptor cells of the retina.[3] This is an organization not only in space but in reaction times.[4]

C. E. Giraldo Herrera, *Microbes and Other Shamanic Beings*,
https://doi.org/10.1007/978-3-319-71318-2_7

In spite of all its refinement, the organization of the vertebrate retina[5] has some equally perplexing, apparently 'flawed', features, which are further aggravated in the case of most mammals, amongst them, humans.[6] For evolutionary biology, these flaws reveal the improvised and accidental character of the eyes' evolution, or alternatively the utter incompetence of any purported 'designer'.[7] First, the wiring awkwardly covers the sensors (see Fig. 9.2): between photoreceptors cells and the light from the outside world there are two neuronal layers, one composed of bipolar cells, horizontal cells, and amacrine cells processing and exchanging information with photoreceptors, and another one of ganglion cells, which process and drive the information to the optical nerve. As the axons of the ganglion cells bundle together, they markedly reduce vision, creating the blind spot of the optical disc. Moreover, these neuronal networks are profusely irrigated by the retinal arteries, veins and capillary vessels fuelling and cleansing the whole system, generating shadows which leave clear traces in a detailed representation of the retinal vessel network in the cerebral cortex.[8] To make matters worse, the photoreceptors are facing backwards, with their body in the path of light making another obstructing layer, and the photosensible tip pointing away from the light sources towards the choroids, another vascular area. Yes, contrary to generalized assumptions (e.g. Wittgenstein and Roy Wagner's Coyote[9]) the eye is within the field of vision, even the photoreceptors (Fig 7.1).

That is, like the prisoners in Plato's 'Cavern',[10] the photoreceptor cells practically look towards the shadows, but not only that, as in the cavern between them and the light source in the pupil, there is a network of roads, through which a myriad caterers and messengers are running. Besides neurons *synapsing*, blood irrigation vessels and capillaries are heavily transited by red and white blood cells, plackets, nutrients, wastes, and occasionally parasite (or perhaps symbiotic) microbes, all of which pass between the receptors and the light source, cluttering vision. Furthermore, like the fire in the cavern, all this electrical activity and movement, heat and pressure are sources of noisy stimuli. All this potentially visible noise and interference arising from things happening in the eye give rise to entoptic phenomena.

To see what happens outside, the image has to be edited by neurons all the way from the retina to the visual cortex, which actively filter and fill

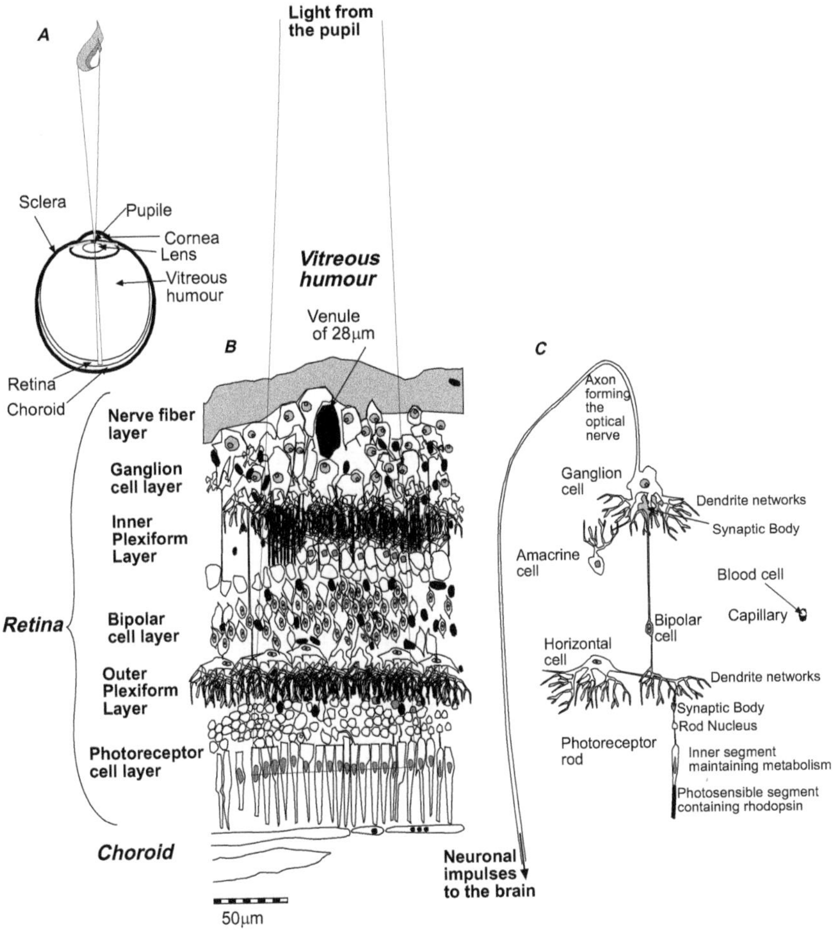

Fig. 7.1 Structure and organization of the retina **(a)** Diagram of the eye; **(b)** Drawing of a micro-section of the primate retina, after a photograph in Adams and Horton (2003); **(c)** Diagram of the kinds of cells that compose the tissue

this dynamic filigree—a highly demanding cognitive task.[11] The arrangement of the eye implies that *exoptic* vision, visual sensibility to events happening outside, is much lower than what would be possible, and that a good deal of resolution is lost in the process. Only when the process of filtering and editing comes to a halt do we become aware of all those things happening in the eye, and we see entoptic phenomena. Under the appropriate lighting and/or metabolic conditions, the filtering process can be overridden, and even without training, some entoptic phenomena are readily accessible, revealing a microscopic world and its underpinnings. In fact, the foundations of our contemporary understanding of the eye and the retina were grounded on entoptic observations.

7.1 Early Physiological Research on Entoptic Vision

Through the Age of Enlightenment until the early twentieth century, entoptic phenomena were explored, described, and articulated into explanations of the organization and workings of the eye. The research on entoptics fructified, while Western medicine, nurtured by botanic and medical knowledge acquired through colonial enterprises and armed with scalpels and microscopes, veered away from humourism, embracing the modern *ontological theory of disease*, which assumes that diseases are caused by entities, which exist on their own independently of the body and its humors.[12]

Some of the figures of this early research, like Boissier de Sauvages, Purkyne, and Marshall were botanist–physicians, with interest and privileged access to information and materials arriving from overseas.[13] Through their experiments, dissections, and microscopy of tissues, these researchers developed an explanation of entoptic phenomena and seated the basis of the modern understanding of the eye. Most of the experiments that constituted this line of research require access neither to expensive equipment nor to exotic or illegal substances.

Boissier de Sauvages (1763b, 238–242) mentions how, by illuminating the white of his eye with a sharp beam of light (transluminating the fundus), he was able to distinguish the palpitating arteries, and the veins of the retina. He also mentions how under inflammation he was able to see *globuli cruoris* (blood globules), like the ones that are visible in the frog's interdigital membrane with the aid of a microscope. He suggested that these phenomena could be behind vertigo and hallucinations.[14]

Half a century later, Purkyne (1819) systematically explored entoptic phenomena as a subjective means to reveal the objective anatomy and physiology of the eye. Purkyne experimented with the effects of illuminating the eyes in diverse ways. Illuminating the normal path of vision with a sharp beam of light (sieving light through a pinhole), he was able to draw detailed depictions of the system of retinal circulatory vessels—which has since been called *Purkyne's tree* (Fig. 10.1: 23–24). He also illustrated blood cells flowing through retinal capillaries (Fig. 10.1: 21 and 29). Purkyne elaborated on the understanding of pressure phosphenes, which are induced by mechanical pressure on the eye, as well as their interference patterns, revealing the neuronal mechanisms that produce the composite images of stereoscopic vision. He also reported on the blue arcs phenomenon, whereby the bundles of neuronal fibres become temporarily perceivable during adaptation to darkness.[15] On the basis of this research, Purkyne grounded a prolific and multifaceted carrier with vast recognition beyond the academia. This success might have been partly due to the fact that at a time when microscopes were still uncommon and expensive equipment and microphotography unimagined, entoptic phenomena served to reify the images drawn from microscopic observation, providing a set of experiences to which laypeople could relate (Fig. 7.2).

Half a century later, Von Helmholtz (1867) proposed means to circumvent the entoptic dampening effect. He showed that employing an oscillating light source, it was possible to stimulate different photoreceptors subsequently, presenting them with the images of entoptic events, thereby avoiding the disappearance of the entoptic image. A physicist–physician Helmholtz also invented ophthalmoscope, an instrument that allows the objective observation of the retinal fundus. Although the ophthalmoscope has far lower resolution than what is afforded subjectively by the methods described by Purkyne, this apparatus permitted the establishment of ophthalmology as a medical specialty,[16] and displaced entoptic vision to the role of an inconvenient curiosity, which nevertheless presented some insights.

For instance, Marshall, an author of manuals of *materia medica* and microscopy, who became interested in entoptic phenomena because of their interference to microscopic observations, elaborated on the observations of

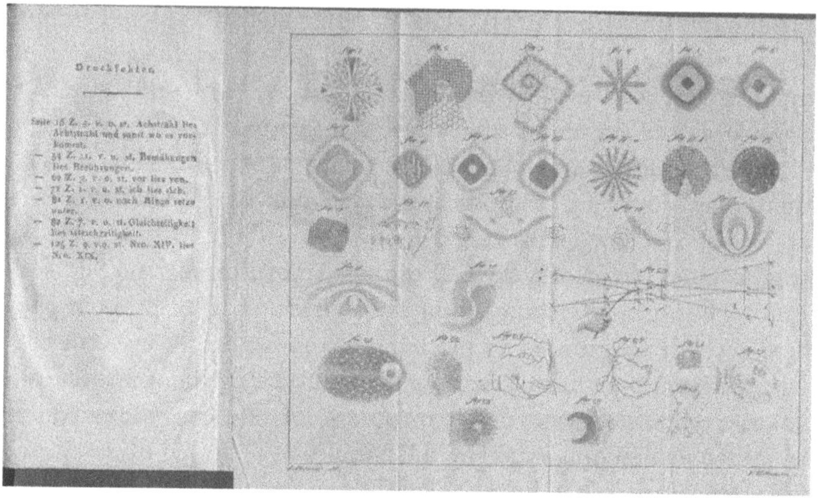

Fig. 7.2 Drawings of various entoptic phenomena (Purkyne 1819) reproduced with the kind permission of the Max Planck Institute for the History of Science. Figs. 1–4 refractive interference patterns under bright illumination; Figs. 21, 26 blood cells; Figs. 23 and 24 'Purkyne's tree' the VESSELS of his eye vascular network

blood corpuscles, through the observation of the light collimated by lenses of the microscope (without slide). He points out that practice increases the degree of observable detail, and blood cells and other structures can in time be seen with different magnifications, often simultaneously.[17]

7.2 Neurophysiology of Eidetic Phosphenes

At the turn of the twentieth century, the rising sciences of neurology, neurochemistry, and psychology emphasized the exploration of hallucinations from a different angle. Klüver (1926), a psychologist and neurophysiologist who developed multiple experiments with mescaline, described how through these experiments it was possible to experience distortions of time and space, an increase in perceived luminosity, and a series of phenomena that he would later call 'constant-forms'. He related these experiences to 'eidetic visions' described under certain neurological conditions related to photographic memory and presumed they were neurogenic (of neural origin). Klüver assumed that mescaline mainly interfered with neuronal communication, revealing the underlying structures of neural function.

A few years later, Marshall[18] carried out his own experiments with mescaline, which would lead him to identify 'eidetic visions' as entoptic phenomena. Marshall confirmed previous observations about how in the early stages of mescaline intoxication, there was a pronounced increase of sensibility to luminosity and colour perception, along with changes in the perception of temporality, and indicated that the observation of entoptic phenomena was facilitated even far below the normal minimal luminosity threshold for these phenomena. Considering the generalized augmented luminosity threshold, he offered entoptic phenomena as a physiological explanation for the constant forms described by Klüver.

Supposedly, Klüver rejected the entoptic explanation concluding that mescaline and eidetic visions 'had form and therefore were not simply ocular dust'.[19] Although Klüver (1942, 177) starts his paper with a similar statement, later on he acknowledges Marshall's points, recognizing that Purkyne could observe 'constant forms' through entoptic vision. Through most of the paper, he attempted to conciliate neurogenic and entoptic views, arguing that as complex phenomena, hallucinations have multiple aspects, which must be taken into account.

However, the neurogenic theory received a boost from neurophysiological research,[20] indicating that by electrically stimulating the open cortex during surgery, it was possible to elicit phenomena resembling pressure phosphenes (brilliant patterns that become apparent while applying pressure on the eyeballs, mechanically stimulating photoreceptors on the retina). To some researchers,[21] the possibility of eliciting these phenomena by stimulating the cortex demonstrated that these and other phosphenes originated in the cortex, rather than in the retina. However, direct electrical stimulation of the brain is poorly understood; it produces multiple effects, which are actually difficult to replicate or isolate.[22] Extraneous stimulation of the cortex could also emulate signals of retinal origin or disinhibit the perception of signals arriving from the retina.[23]

Neurophysiological research continued to flourish, demonstrating that phosphenes could also be induced by electrical currents of magnitudes and frequencies simulating electroencephalographic signals, with electrodes placed over the skin of skull, without surgery. Through these means it is possible to elicit some 15 different phosphene shapes correlated with particular bandwidths of electromagnetic stimulation.[24] This phenomenon can also be elicited through the magnetic stimulation of either the eyes or the surface of the skull, indicating the sensibility of either photoreceptors or neurons to electromagnetic stimulus.[25] Nevertheless, as Knoll points out, these means are inadequate to stimulate the cortex directly, they require a resonance mechanism linked specifically to the visual system, and it is likely that the retinal system is being stimulated directly. Thus, the question remains of whether the images are being produced by photoreceptors of the retina or somewhere along the pathway to the cortex.[26]

Transcranial stimulation has been greatly refined with a repertoire of methods which employ magnetic, constant, and alternate electrical current, that offer greater possibilities to focus on specific areas of the visual cortex. In this way, it has been possible to elicit more specific phosphene responses, even in 'blind' people.[27] However, these new approaches fail to address the objections raised against the earlier experiments. 'Blind' individuals actually presented residual vision, or damages involving the cornea rather than the retina, and the authors recognize that the definition of their phosphenes was poor. The researchers failed to examine whether phosphenes could be also elicited through such electromagnetic stimulation applied beyond the visual cortex.[28] Thus, the issue of whether electromagnetically induced phosphenes are neurogenic or optic phenomena continues to be the subject of heated debates in physiology.

7.3 Pharmacology of Hallucinations and Phosphenes

On the pharmacological level, the issue of the origin of hallucinations and phosphenes does not present an easier solution. The mechanisms whereby serotonergic hallucinogens induce perceptual distortions remain to be elucidated. Serotonin and its agonists act at multiple levels of the brain. However, they do not act directly over the visual cortex. On the cortex, their action is concentrated on the prefrontal area, which regulates personality, attention, memory, learning, and sociability. On the brainstem, serotonin and its agonists also block spontaneous activity of the locus coeruleus, which regulates stress responses, and simultaneously increases its sensitivity to external stimuli, inducing an attentive walking state, in which external stimuli come to be regarded as novel and relevant.[29] This would suggest that the perceptual distortions evocated by serotonergic substances and classified as hallucinations may arise from alterations in the integration and valuation of objective stimuli rather than from spontaneous brain activity.

While substances inducing hallucinations are closely related and linked to the serotonergic system, the substances that induce phosphenes are extremely varied in their molecular structure, in their physiological and psychiatric effects. In fact, the induction of phosphenes is one of the most frequent side effects of drugs. The precise mechanisms whereby phosphenes are induced are poorly understood for most substances. It is possible that some substances operate directly either stimulating or disinhibiting photoreceptors in the retina, others could be stimulating neurons along the visual cortex, or yet halting the dampening mechanisms. The pharmacological induction of phosphenes by ivabradine (an anti-anginal drug) can be explained by the simultaneous hypersensitization of photoreceptors and the inhibition of the filtering mechanisms dampening slow responses, exposing the system to higher levels of noise.[30] Similar mechanisms seemingly operate over the induction of phosphenes by serotonin and its agonists.[31]

Although yage, yopo, and various other substances employed by some Lowland South American shamans contain serotonergic hallucinogens, the mechanisms whereby serotonergic hallucinogens produce hallucinations remain to be elucidated. What has been demonstrated so far is that these substances enhance attention to stimuli from the sensorium. On the other hand, the pharmacological induction of phosphenes seemingly operate at the retinal level augmenting the visual threshold exposing the system to higher levels of 'noise', allowing the registration of entoptic images. For this reason, it is convenient to explore the insights of recent developments of research on entoptics.

7.4 Renewed Interest for Entoptics

Although largely displaced by apparatus in ophthalmology and neglected or ignored by psychologists and anthropologists, research on entoptics as optical phenomena has witnessed a revival in the last 40 years. Boissier de Sauvage, Purkyne, Helmholz, and Marshall noted how, with some training, it is possible to distinguish the synchronous movement of the blood corpuscles rhythmically circulating through the capillaries of the retina, accelerating and pausing with the pulse. Since the 1980s, this phenomenon has been clarified and its observation further refined, into what today is known as the *blue field effect*. Bright light, in particular the blue light, allows the observation of 7–15 μm in diameter, white blood cells, refracting light like microscopic lenses. Erythrocytes, red blood cells, are not visible under these conditions because they absorb rather than refract light of this wavelength, and the wavelengths they refract are a peak of low sensitivity for photoreceptor molecules. Their contours, however, can be apparent and occasionally their individual shadows. The blue field effect, although subjective, is still one of the best means to assess the status of the immune system, as well as blood flow and speed, in particular in the eye where until recently, other means were not available or are extremely expensive.[32]

In a similar fashion, Helmholz's destabilizing technique was further refined, employing different oscillatory frequencies to sharpen the shadow images of layers of vessels at different distances from the retina. In order to see objects close to the retina, the light source must move fast, while to see objects far from the retina, the light source must move more slowly. The identity of vessels and capillaries seen entoptically has been confirmed in histological preparations and through angiographic fluorescence photography.[33] These arrangements allow for entoptic observations of a quality that can be assessed, so as to compare the correspondence between the images of the vessel network drawn from fluorescence photography with those drawn from high contrast entoptic images. Entoptic images have higher resolution than fluorescence photography. The technique is inexpensive and has been useful in the diagnostic control of diabetes and other diseases affecting the retina, such as those caused by systemic infection.[34] Objects of known sizes, like capillaries or blood cells, are visible, but it is evident that more meets the eye than would be expected. Smaller structures are visible, even though they elude corrosive histological preparations.[35]

Entoptic vision allows us to witness physical events happening within the eye. It has been one of the fundaments of Western understanding of the structure and the workings of the eye. Now, techniques that enhance entoptic vision are being developed into diagnostic tools by biomedicine. Some of these techniques produce similar effects to those employed in the course of shamanic practices. For example, the light-effects of Helmholz's destabilizing techniques resemble those produced by the complex rhythmic movements of the *waira-sacha* fan-leaves, filtering light through its pinholes.

Perhaps, as anthropologists we should consider what are the epistemological implications of being able to look within our own selves? What does it mean for shamans to have phenomenological access to the microscopic world flowing through their veins?

Notes

1. Pirenne 1953; Denton and Pirenne 1954; Baylor 1996; Baylor et al. 1979, 1984.
2. Cronin 2008.
3. Curcio et al. 1990.
4. Stockman et al. 2010.
5. Pough et al. 1996.
6. Walls 1991, 476.
7. Dawkins 1996, 93.
8. Adams and Horton 2002, 2003.
9. Wagner 2010, 4 & 48; Wittgenstein 2017 [1922], 5.331. This implies a very different interpretation of what *nagual, second attention* and the exercises Don Juan played on Castaneda than what is delivered by Wagner.
10. Shorey 1969, 7.514a.
11. Coppola and Purves 1996.
12. Rosen 1944.

13. Schiebinger 2009; Wade et al. 2001; Ramawat 2008; Altman 1987.
14. Boissier de Sauvages 1763a, 238–242.
15. Moreland 1968.
16. Rosen 1944.
17. Marshall 1935.
18. Marshall 1937.
19. In Lewis-Williams et al. 1988, 202.
20. Review in Brindley 1973.
21. Hodgson 2000; Brindley 1973.
22. Borchers et al. 2012.
23. Schwiedrzik 2009.
24. Knoll et al. 1963.
25. Marg and Rudiak 1994.
26. Knoll et al. 1963, 217.
27. Fernandez et al. 2002; Kanai et al. 2008.
28. Schwiedrzik 2009.
29. Nichols 2004; Aghajanian and Marek 1999.
30. Cervetto et al. 2007.
31. Gu 2007.
32. Sinclair et al. 1989.
33. Applegate et al. 1990; Bird and Wealer 1974.
34. Bradley et al. 1998.
35. Applegate et al. 1990; Bird and Wealer 1974.

Bibliography

Adams, D.L., and J.C. Horton. 2002. Shadows cast by retinal blood vessels mapped in primary visual cortex. *Science* 298 (5593): 572.

———. 2003. The representation of retinal blood vessels in primate striate cortex. *Journal of Neuroscience* 23 (14): 5984.

Aghajanian, G.K., and G.J. Marek. 1999. Serotonin and hallucinogens. *Neuropsychopharmacology* 21 (S1): 16S–23S.

Altman, L.K. 1987. *Who goes first?: The story of self-experimentation in medicine.* Berkeley: University of California Press.

Applegate, R.A., A. Bradley, and W.A. van Heuven. 1990. Entoptic visualization of the retinal vasculature near fixation. *Investigative Ophthalmology and Visual Science* 31 (10): 2088–2098.

Baylor, D. 1996. How photons start vision. *Proceedings of the National Academy of Sciences of the United States of America* 93 (2): 560–565.

Baylor, D.A., T.D. Lamb, and K.W. Yau. 1979. Responses of retinal rods to single photons. *The Journal of Physiology* 288 (1): 613–634.

Baylor, D.A., B.J. Nunn, and J.L. Schnapf. 1984. The photocurrent, noise and spectral sensitivity of rods of the monkey *Macaca fascicularis*. *Journal of Physiology* 357 (1): 575–607.

Bird, A.C., and R.A. Wealer. 1974. On the retinal vasculature of the human fovea. *Experimental Eye Research* 19 (5): 409–417.

Boissier de Sauvages, François. 1763a. *Nosologia methodica sistems morborum classes, genera et species, Juxtà Sydenhami mentem et Botanicorum ordinem. Tomi tertii pars prima.* Amsterdam: Frères De Tournes; (c) Bibliothèque interuniversitaire de médecine (Paris).

———. 1763b. *Methodical system of disease nosology*, vol. 3. Amsterdam: Frères De Tournes.

Borchers, Svenja, Marc Himmelbach, Nikos Logothetis, and Hans-Otto Karnath. 2012. Direct electrical stimulation of human cortex – the gold standard for mapping brain functions? *Nature Reviews Neuroscience* 13 (1): 63–70.

Bradley, A., H. Zhang, R.A. Applegate, L.N. Thibos, and A.E. Elsner. 1998. Entoptic image quality of the retinal vasculature. *Vision Research* 38 (17): 2685–2696.

Brindley, Giles S. 1973. Sensory effects of electrical stimulation of the visual and paravisual cortex in man. In *Visual centers in the brain*, 583–594. Berlin: Springer.

Cervetto, L., G.C. Demontis, and C. Gargini. 2007. Cellular mechanisms underlying the pharmacological induction of phosphenes. *British Journal of Pharmacology* 150 (4): 383–390.

Coppola, D., and D. Purves. 1996. The extraordinarily rapid disappearance of entopic images. *Proceedings of the National Academy of Sciences of the United States of America* 93 (15): 8001–8004.

Cronin, T.W. 2008. Visual ecology. In *The senses: A comprehensive reference*, ed. R. Masland and T.D. Albright, 211–245. New York: Academic Press.

Curcio, Christine A., Kenneth R. Sloan, Robert E. Kalina, and Anita E. Hendrickson. 1990. Human photoreceptor topography. *Journal of Comparative Neurology* 292 (4): 497–523.

Dawkins, R. 1996. *The blind watchmaker: Why the evidence of evolution reveals a universe without design*. New York: WW Norton & Company.

Denton, E.J., and M.H. Pirenne. 1954. The absolute sensitivity and functional stability of the human eye. *Journal of Physiology* 123 (3): 417–442.

Fernandez, E., A. Alfaro, J.M. Tormos, R. Climent, M. Martínez, H. Vilanova, V. Walsh, and A. Pascual-Leone. 2002. Mapping of the human visual cortex using image-guided transcranial magnetic stimulation. *Brain Research Protocols* 10 (2): 115–124.

Gu, Qiang. 2007. *Serotonin involvement in plasticity of the visual cortex. In Monoaminergic modulation of cortical excitability*, 113–124. Boston: Springer.

Hodgson, Derek. 2000. Shamanism, phosphenes, and early art: An alternative synthesis. *Current Anthropology* 41 (5): 866–873. https://doi.org/10.1086/317415.

Kanai, Ryota, Leila Chaieb, Andrea Antal, Vincent Walsh, and Walter Paulus. 2008. Frequency-dependent electrical stimulation of the visual cortex. *Current Biology* 18 (23): 1839–1843.

Klüver, Heinrich. 1926. Mescal visions and eidetic vision. *The American Journal of Psychology* 37 (4): 502–515.

———. 1942. Mechanisms of hallucinations. In *Studies in personality*, 175–207. New York: McGraw-Hill.

Knoll, M., J. Kugler, O. Hofer, and S.D. Lawder. 1963. Effects of chemical stimulation of electrically-induced phosphenes on their bandwidth, shape, number and intensity. *Confinia Neurologica* 23 (3): 201–226.

Lewis-Williams, J.D., T.A. Dowson, Paul G. Bahn, H.G. Bandi, Robert G. Bednarik, John Clegg, Mario Consens, Whitney Davis, Brigitte Delluc, Gilles Delluc, Paul Faulstich, John Halverson, Robert Layton, Colin Martindale, Vil Mirimanov, Christy G. Turner, Joan M. Vastokas, Michael Winkelman, and Wylie Alison. 1988. The signs of all times: Entoptic phenomena in upper palaeolithic art [and comments and reply]. *Current Anthropology* 29 (2): 201–245.

Marg, Elwin, and David Rudiak. 1994. Phosphenes induced by magnetic stimulation over the occipital brain: Description and probable site of stimulation. *Optometry and Vision Science* 71 (5): 301–311.

Marshall, Charles Robershaw. 1935. Entoptic phenomena associated with the retina. *British Medical Journal* 19 (4): 177.

———. 1937. An enquiry into the causes of mescal visions. *The Journal of Neurology and Psychopathology* 1 (68): 289.

Moreland, J.D. 1968. Threshold measurements of the blue arcs phenomenon. *Vision Research* 8 (8): 1093–1106.

Nichols, David E. 2004. Hallucinogens. *Pharmacology & Therapeutics* 101 (2): 131–181. https://doi.org/10.1016/j.pharmthera.2003.11.002.

Pirenne, M.H. 1953. The absolute sensitivity of the eye and the variation of visual acuity with intensity. *British Medical Bulletin* 9 (1): 61.

Pough, Harvey, John Heiser, and William N. Macfarland. 1996. *Vertebrate life*. 4th ed. Upper Saddle River: Prentice Hall.

Purkyne, Jan Evangelista. 1819. *Beiträge zur Kenntniss des Sehens in subjectiver Hinsicht*. Prague: Edlen von Wildenbrunn, European Cultural Heritage Online.

Ramawat, K.G. 2008. *Herbal drugs: Ethnomedicine to modern medicine*. Berlin: Springer Berlin Heidelberg.

Rosen, G. 1944. *The specialization of medicine: With particular reference to ophthalmology*. New York: Froben.

Schwiedrzik, C.M. 2009. Retina or visual cortex? The site of phosphene induction by transcranial alternating current stimulation. *Frontiers in Integrative Neuroscience* 3 (6): 1–2. https://doi.org/10.3389/neuro.07.006.2009.

Schiebinger, Londa L. 2009. *Plants and empire*. Cambridge, MA: Harvard University Press.

Shorey, Paul, ed. 1969. *Republic, plato in twelve volumes*, Vols. 5 & 6. Trans. Paul Shorey. Cambride, MA: Harvard University Press; William Heinemann, Perseus Digital Library Project. ed. Gregory R. Crane. Tufts University.

Sinclair, S.H., M. Azar-Cavanagh, K.A. Soper, R.F. Tuma, and H.N. Mayrovitz. 1989. Investigation of the source of the blue field entoptic phenomenon. *Investigative Ophthalmology and Visual Science* 30 (4): 668.

Stockman, Andrew, Toby Candler, and Lindsay T. Sharpe. 2010. Human scotopic sensitivity is regulated postreceptorally by changing the speed of the scotopic response. *Journal of Vision* 10 (2): 1–19. https://doi.org/10.1167/10.2.12.

Von Helmholtz, H. 1867. *Handbuch der physiologischen Optik*. Leipzig: Voss.

Wade, N.J., J. Brozek, and J. Hoskovec. 2001. *Purkinje's vision: The dawning of neuroscience*. Oxford: Taylor & Francis.

Wagner, Roy. 2010. *Coyote anthropology*. Lincoln/London: University of Nebraska Press.

Walls, Gordon L. 1991. Appendix: Mammals. In *Evolution of the eye and visual system*, ed. J. Cronly-Dillon and R.L. Gregory, 467–483. London: Macmillan.

Wittgenstein, Ludwig. 2017 [1922]. *Tractatus Logico-Philosophicus / Logisch-philosophische Abhandlung, side-by-side edition containing the original German, alongside both the English translations. Version 0.45*. Trans. Ogden and Ramsey / Pears and McGuinness. London: Kegan Paul. Kevin Klement. http://people.umass.edu/klement/tlp/.

8

Entoptic Microscopy

Before answering what are the epistemological consequences of entoptic microscopy for our understanding of shamanism, it is necessary to determine whether this form of vision corresponds to what shamans see. To answer this question, we will analyze what characterizes entoptic microscopy images? What differentiates them from purportedly neurogenic images? And whether, and to what degree, the characteristics of either kind of image correspond to the depictions of shamans. Furthermore, we will inquire what might be visible of the world within. What resolution can we expect of this microscopy?

8.1 Characteristic Features of Entoptic Images

Strictly speaking, most entoptic images are shadows and refractions of light, and as such they are characterized by geometrical features which result from the relation of the sizes of the light source and the object casting a shadow, their distance, and the distance to the surface on which the

C. E. Giraldo Herrera, *Microbes and Other Shamanic Beings*,
https://doi.org/10.1007/978-3-319-71318-2_8

image is being projected. If the light source area is equal to that of the object, light rays are parallel and the shadow will be of equal size.

However, the diameter of the pupil is generally larger than objects in the eye, and most corpuscles and capillaries are smaller or thinner than light sources; hence, their shadows better resemble those produced during solar eclipses (see Fig. 7.2). In these circumstances, the shadow is not sharp, and it can be composed of one or two of three regions: an umbral cone and a penumbra, only a penumbra, or a penumbra and an antumbra.

The umbral cone is the region completely obscured by the object; its length depends on the size of the source, the size of the object, and their distance. The umbral cone is surrounded by a penumbral region, partially obscured by the object but illuminated by the exceeding light from the source. Beyond the length of the umbral cone the exceeding light overlaps, creating an antumbra, a shadow more illuminated than the penumbra. In this way, the kind of shadow cast on a surface like the retina will depend on the distance between the object and the surface (retina), and the relation of sizes between light source and object. An object within umbral cone distance will project a shadow with an umbral core and a penumbra, and as distance increases, the umbra will be smaller and the penumbra will grow larger and dimmer, until the former disappears and is replaced by lighter central antumbra. A homologous effect occurs in refracted images.

Jacanamijoy's paintings, like the paintings of other Amerindian and Ayahuasca artists, and the depictions of shamans, illustrate this kind of effect (see Fig. 8.1). The motifs portrayed are presented at various scales. However, increases in size are coupled with an increase in transparency and a decrease in resolution: small whitish commas with clearly delineated outlines, portrayed in slightly larger format, acquire a reddish core. Still larger images of the commas become more and more diffuse, until they become frames for smaller commas and their cores vanish in what seems to be a vague antumbra with a reddish outline. A similar thing occurs with yellow marks, which become empty yellow circles revealing a blue, green, or brown background. As they become larger, they also turn more diffuse. There are also green and blue lines, which, in the smaller version, are painted in highly contrasting colours, and in larger versions

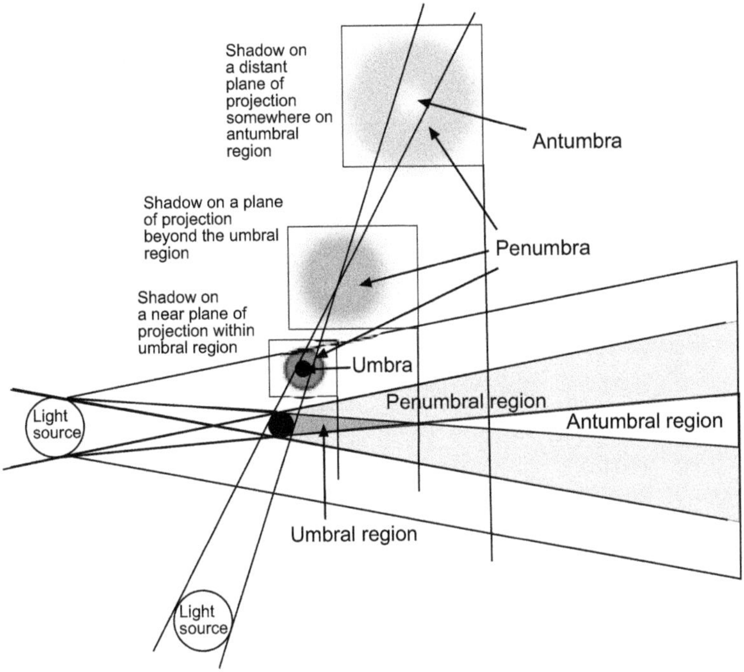

Fig. 8.1 Geometry of shadow projection and formation, a two-dimensional silhouette of the different illuminated regions. The boxes portray the shadow projected at corresponding distances

display larger and larger penumbral shadows. Jacanamijoy's painting reproduces realistically what would be expected of optic phenomena occurring within the eye.

Now if we assume that these images are the optical effects of an object casting shadows or refractive images on the retina, the question that remains is what these objects might be. Some of the motifs in Jacanamijoy's works could represent white blood cells and the vascular network. However, Jacanamijoy portrays a wide variety of motifs. Moreover, there are ethnic specific motifs. If all humans, in fact all vertebrates, share similar retinas with similar blood cells flowing through its vasculature and innervation, why should there by ethnic-specific motifs?

There are many more known entoptic phenomena and many ways of seeing them, which might explain some of these motifs. However, a more

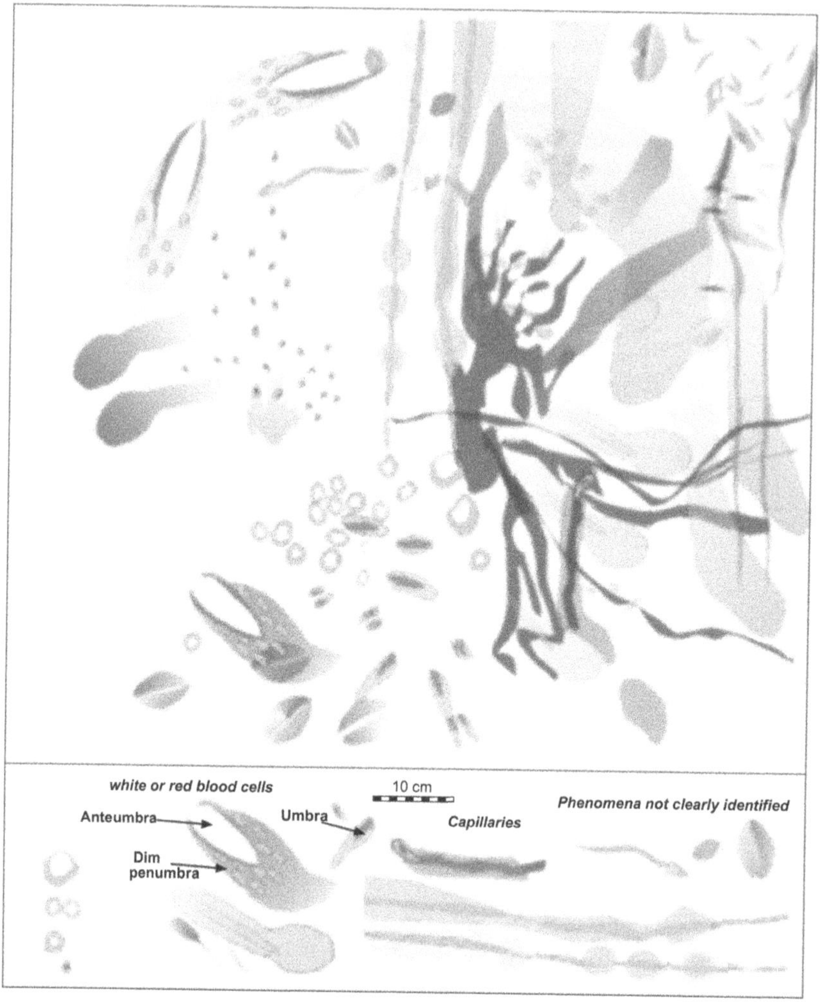

Fig. 8.2 Analytic drawing of Jacanamijoys's *internal navigation* (2009). (**a**) Drawing, (**b**) elements extracted

interesting possibility is that these motifs represent microbes flowing through the vasculature of the retina. This would be in closer agreement with the ontological claims of shamans. To address this possibility, it is necessary to estimate what might be visible through entoptic microscopy (Fig. 8.2).

8.2 What Might Be Visible to Shamans Through Entoptic Microscopy

Entoptic resolution depends on two factors: visual acuity and shadow formation geometry. Visual acuity depends on the adaptation and status of the photoreceptors and the optical nerve, as well as on the attention and practice of the observer. As has been shown, many of the substances employed in lowland South American shamanism propitiate the hypersensitation of photoreceptors and optimize neuronal processes enhancing the attention of the observer. On the other hand, developing the ability to see visions requires extended practice from shamans.[1]

Visual acuity also varies according to radiance (brilliance), wavelength (colour), accommodation (focus), and the shape of the object. As has been seen, some shamanic contexts are illuminated by intense red fires. Red light normally would not be considered optimal for entoptic observation, because the low energy of the wavelength could fail to stimulate cones, the photoreceptors involved in the perception of colour and detail. Nevertheless, this wavelength of light would prevent bleaching hypersensitized photoreceptors. Further experimental research is necessary to address the issue.

Regarding shape, under optimal conditions, for a wavelength of 550 nm (blue light) a healthy eye can resolve 1′ minute of angular distance, that is, it is possible to recognize a standard block letter E of 8.9 mm at 6.096 m; below this threshold, the image becomes increasingly blurred. As a rule of thumb, with the eye focused at infinity, 1′ of the visual field equals to 6 μm at the surface of the retina.[2] The *nonius*, the separation and resolution of two short parallel straight lines, reaches an angle of 10″ seconds of the visual field (1 μm). Further acuity (5″ seconds) is necessary and achieved through stereoscopy to resolve depth.[3] This suggests that the potential resolution of entoptic microscopy is not insignificant; it would be sufficient to observe small protozoa and bacteria. Visual acuity, however, is only half of the problem; we must also consider how shadow formation geometry defines the dimensions and characteristics of the images projected on the retina. This can be achieved taking into consideration the geometry of similar triangles.[4]

Shadow geometry depends on the relation of sizes between the light source and the object casting a shadow, their mutual distance, and the distance to the surface on which the image is being projected. The first vascular layer is at least 80 μm from the photosensible extreme of photoreceptors; this gives us a basic distance to the surface where the image is being projected. Serotonin analogues, like those in yage, dilate the pupil.[5] With a dilated pupil (4 mm), only the largest vessels (65 μm) cast clearly defined shadows beyond 80 μm; capillary shadows are enlarged but become too dim even at this threshold. However, pinholes like those in a *waira-sacha* fan, refractions on a *waira-wawa* crystal or even half-shut eyelids, constitute sharper and more distant light sources, allowing the observation of smaller objects. Illuminated by pinholes of 0.5 mm, at a distance of 3 cm, an object of 1 μm of width, flowing through the capillaries at 80 μm, will cast a penumbral shadow of 1.8 μm, which could fall within the range of human visual acuity. This implies that even bacteria could fall within the range of entoptic microscopy.

Treponema pallidum, the causal agent of syphilis, is about 15 μm in length with a diameter of 1 μm. Protozoan parasites are generally larger, and many like *Leishmania* spp., *Trypanosoma* spp., *Plasmodium* spp., and *Toxoplasma* spp. are involved in tropical zoonotic medical conditions and affections. For long, blood was considered as a sterile environment for most microbes; hence, the presence of microbes in blood is rare, except in tropical contexts, and indicates a serious systemic infection. One of the first organs affected is the eye; hence, ophthalmological lesions are an index of systemic infection.[6] What this suggests is that entoptic vision might be a powerful tool revealing a world within; a world otherwise invisible to the untrained eye without the aid of a microscope and someone skillful to handle it.

Smaller objects might be visible under particular circumstances, in particular if we consider bioluminescence.[7] However, it is highly unlikely that seen through entoptic vision, objects like DNA resemble the theoretical models of molecular biology, as Narby (1999) assumes, without considering the optical aspects of this phenomenon or other entoptic objects with similar appearance, for instance, blood vessels, some blood parasites, and some pathogenic bacteria.

8.3 Seeing Things Together

Lowland South American Amerindian communities recognize the visions of hallucinations and dreams as means to enquire into the fundaments of physical reality. Their ontologies are grounded on these experiences, and their shamans derive from them some of their diagnostic means to understand the body and deal with the pathogenic owners of wildlife. Labelling these experiences as metaphorical constructs or as neurogenic 'short-circuits', symbolic and neuropsychological approaches void these experiences of their empirical value and fail to take shamanism and animic ontologies seriously.

It is possible and desirable to acknowledge animic ontologies without relinquishing the insights developed by natural sciences. As illustrated by the rivalling optic and neurogenic theories of phosphenes, natural sciences are the scenery of profound debates between different ways of substantiating and constituting reality. Thus, they do not present a monolithic ontology. Some of these approaches, like the research on entoptics, can ground the epistemologies of shamanism. Moreover, as evidenced by the development of research on hallucinations, entoptics, and phosphenes, the insights of Western science have not been developed in isolation but rather in contact with non-Western ontologies. Thus, a more profound form of decolonization of thought involves reclaiming natural sciences and their insights.

Physiological insights into entoptic vision reveal it as a form of microscopy afforded by the architecture of the vertebrate retina. Physiological research demonstrated that entoptic microscopy allows perceptual access to retinal objects like veins, arteries, and capillaries, as well as to individual cells like leukocytes and debris flowing in or near the retina. Current ophthalmology employs these insights in diagnosis. It has been suggested that smaller structures might be visible as well. Taking into consideration human visual acuity and the geometry of shadow formation, entoptic microscopy can be estimated to be sufficient to resolve objects of 1 μm, which is the size of bacteria and small protozoa, and the threshold of detection might be smaller still. The retina is highly exposed during systemic infection and various products of microbial metabolism trigger the disinhibition of entoptic vision.

Given that *yage, yakoahani, yopo, peyotl,* and other 'psychotropic' substances employed by shamans augment sensibility to light and enhance the attention of the observers, I hypothesize that shamans have developed enhanced techniques of entoptic microscopy. The usage of brilliant red lights, which optimize visual cell sensitivity of powder, crystals, and fans that filter and concentrate light sources and through their movement destabilize images, avoiding the dampening responses and controlling resolution, corroborates this hypothesis. The hypothesis is substantiated by the presence of the characteristic features derived of shadow geometry, which indicate the optical origin of some of these visions and provide means for the measurement of entoptic objects. Such image characteristics feature in the paintings by Jacanamijoy of his experiences with *yage*. Furthermore, some of the images in these paintings correspond with entoptically visible microscopic structures, including capillaries and individual blood cells flowing through them, while other elements remain unidentified, suggesting the need for closer collaborative experimental research. The possibility of observing cellular structures and microbes through entoptic microscopy substantiates shamanic epistemologies, further bridging the gap between shamanism and microbial ecology.

Entoptic microscopy affords access to an inner microscopic reality acknowledged by biomedicine. However, it does not exhaust the meaning and content of shamanic visions, not even the relation of these visions with the microbial world. We must bear in mind that, as was mentioned in Chap. 4, the active principles of many of the substances employed by shamans are molecules which microorganisms produce and employ for their communication and interactions, both amongst themselves and with our neuroendocrine system. Other aspects of shamanic visions may be expressions of these paths of communication.

Notes

1. Kopenawa and Albert 2013.
2. Atchison and Smith 2000; Hallett et al. 1962.
3. Gross 2008, 57–63.
4. Adams and Horton 2002; Applegate et al. 1990.

5. Callaway et al. 1999.
6. Bouza et al. 2000; Callegan et al. 2007.
7. Bókkon 2008; Bókkon and Vimal 2009.

Bibliography

Adams, D.L., and J.C. Horton. 2002. Shadows cast by retinal blood vessels mapped in primary visual cortex. *Science* 298 (5593): 572.

Applegate, R.A., A. Bradley, and W.A. van Heuven. 1990. Entoptic visualization of the retinal vasculature near fixation. *Investigative Ophthalmology & Visual Science* 31 (10): 2088–2098.

Atchison, D.A., and G. Smith. 2000. *Optics of the human eye*. Oxford: Butterworth-Heinemann Medical.

Bókkon, István. 2008. Phosphene phenomenon: A new concept. *Biosystems* 92 (2): 168–174.

Bókkon, István, and Ram Lakhan Pandey Vimal. 2009. Retinal phosphenes and discrete dark noises in rods: A new biophysical framework. *Journal of Photochemistry and Photobiology B: Biology* 96 (3): 255–259.

Bouza, Emilio, Rosario Cobo-Soriano, Marta Rodríguez-Créixems, Patricia Muñoz, Marta Suárez-Leoz, and Carlos Cortés. 2000. A prospective search for ocular lesions in hospitalized patients with significant bacteremia. *Clinical Infectious Diseases* 30 (2): 306–312.

Callaway, J.C., D.J. McKenna, C.S. Grob, G.S. Brito, L.P. Raymon, R.E. Poland, E.N. Andrade, E.O. Andrade, and D.C. Mash. 1999. Pharmacokinetics of Hoasca alkaloids in healthy humans. *Journal of Ethnopharmacology* 65 (3): 243–256. https://doi.org/10.1016/S0378-8741(98)00168-8.

Callegan, Michelle C., Michael S. Gilmore, Meredith Gregory, Raniyah T. Ramadan, Brandt J. Wiskur, Andrea L. Moyer, Jonathan J. Hunt, and Billy D. Novosad. 2007. Bacterial endophthalmitis: Therapeutic challenges and host-pathogen interactions. *Progress in Retinal and Eye Research* 26 (2): 189–203.

Gross, H., ed. 2008. *Handbook of optical systems vol. 4 survey of optical instruments*. Weinheim: Wiley-VCH Verlag GmbH, Wiley Online Library.

Hallett, P.E., F.H.C. Marriott, and F.C. Rodger. 1962. The relationship of visual threshold to retinal position and area. *Journal of Physiology* 160 (2): 364–373.

Kopenawa, Davi, and Bruce Albert. 2013. *The falling sky: Words of a Yanomami Shaman*. Trans. Nicholas Elliott and Alison Dundy. Cambridge, MA: Harvard University Press.

Narby, Jeremy. 1999. *The cosmic serpent: DNA and the origins of knowledge*. New York: Tarcher; Putnam.

Part III

Biocultural Ethnohistory of the Zemes of Treponematoses

In Chap. 3, we saw how Amerinds in the Caribbean during the sixteenth and seventeenth centuries regarded *Zemes* and other *Acançancóu* or *Oúpoyem,* the category of entities friars translated as spirits, as perceivable beings, which required nourishment and were susceptible to material substances, aged, reproduced, proliferated, recognized their kin, affected the behaviour, and protected or promoted certain places, animals, plants, communities, and individuals, and caused diseases. In Chap. 4, we saw how these understandings are shared by current shamans and their partial commensurability with current microbiological understandings. In the Second Part, we explored how the forms of perception through which shamans engage with these entities may grant them access to the microbial world, and how the images derived from entoptic microscopy correspond to their depictions of the world of shamanic beings. To address whether microbiology finds microorganisms where Amerind shamans perceive shamanic beings, the Third Part returns to examine sixteenth and seventeenth-century records of some zemes that were the main characters of the myth of *The Sun and the Moon,* and which were identified with the French malaise or the pox, that is, syphilis. The analysis of the myths, and its correlation with archaeological findings, corroborate this identification and reveal further references to this and other treponemal diseases, to their understanding, their treatments, and their importance in Amerind societies. We examine how Europeans came to share the realities of these diseases and how Amerind knowledge about them was purified, translated, and adapted into European medicine. I argue that this translation served as an ontological scaffold for the development of microbiology.

The West holds contradictory stances towards myths. Myths are often dismissed as naïve explanations of natural phenomena, as rough precursors of religion and science, as fanciful accounts of history, or as expressions of obscure aspects of the mind. Myths are, nonetheless, also cherished for their symbolic wealth and their ever-defying imaginative resourcefulness. Myths are rarely, if ever, simple or obvious. In their intricacy, they might seem childish and sometimes even jocosely absurd. However, myths are multilayered. They talk in different levels to different people. They thread moral and social norms with specific allusions to astronomy, features of the landscape, kinship, flora, fauna, technologies, and particular events in history.[1]

The symbolic wealth of myths does not arise from fanciful delusion but from intricate and intimate relations with the lived world.[2] The observations about ecology and behaviour in myths are often demonstrated to be the accurate product of excellent naturalists.[3] Thus, myths can be seen like equations, in which symbols have become heavily loaded, registering, relating, and elaborating on observations carried by different peoples through various generations. Lévi-Strauss noted how Amerindian myths reflect social relations and how their transformations reflect the relations of different communities to one another.[4] Given that Amerindian ontologies do not restrict the social community to the human species, myths also constitute an important and insightful record that provides keys about nature and the way it has been constituted through the social interactions between humans and non-humans. Moreover, many Amerind groups understand myths as narratives about the world of shamanic beings.[5]

This part explores a synthetic path that understands myth and related knowledge as records of a biosocial or socio-natural history. It will explore this often-neglected layer of interpretation, focusing on myths as records of masters of game and other shamanic beings, and the diseases with which they were associated. It will address the social interactions between and beyond humans portrayed by Taino-Callinago and Nahuatl on early accounts of the myth of *The Sun and the Moon*, analysing the understandings they contain about treponemal diseases, with which they were explicitly associated. Moreover, this part will show how the interactions with these entities could have influenced the development and evolution of syphilis and closely related diseases. It also explores how biological understandings of the disease may contribute to the interpretation of the meaning and symbolism in myth. This synthetic approach could, perhaps, be assimilated to naturalistic medical materialism, but only insofar as the latter approached Amerindian perspectivism, acknowledging the importance of sociality and subjectivity beyond humans.

Syphilis, yaws, and pinta are closely related diseases that feature prominently in some of the earliest accounts of Amerindian medical knowledge and mythology. These diseases, the way they were understood by Amerindians, and their treatments drew the avid attention of European missionaries, chroniclers, and historians in the years that followed the first contacts. According to these records and to the oral traditions of some enduring communities, these diseases were starring characters of their myths origin. They are protagonists of the earliest recorded versions of the myth of the Sun and the Moon—a myth, which, albeit with profound variations, is widely distributed throughout the Americas. This myth narrates the events that led to the origin of the celestial bodies and their spots, of crucial cultural practices like fishing, cultivation, pottery or metallurgy, and of diseases related with syphilis.

These records have been mentioned, tangentially, in the heated controversies that have surrounded the origin of syphilis.[6] As will be seen later on, these controversies have been largely settled, suggesting an Amerindian origin, with possible subsequent mutations after the transmission to the old world. [7] However, beyond these controversies, what is readily apparent is that Amerindians not only suffered these pathologies, but also developed insightful understandings of them, which coetaneous Europeans appreciated highly. Amerindian medical knowledge and even the mythology associated with it were quickly assimilated into Western medical treatises dealing with what at the time appeared like an emerging disease.[8] One of the physicians who conducted this assimilation is the earliest proposer of the theory of contagion in the West.

As in other instances,[9] through the process of assimilation, Amerindian knowledge was dissociated and purified into descriptions amenable to naturalist explanation, or into myths and antiquities seemingly more susceptible to theological or humanist exegesis. Nevertheless, after the division of the spoils, some of the knowledge that was deemed as mythological ended up being rediscovered by medicine, and some of the knowledge that was initially recognized as factual fell into disbelief.

Notes

1. Lévi-Strauss 1958, 1995; Basso 2009; Santos-Granero 1998.
2. Malinowski 1948 [1926].
3. Lévi-Strauss 1982, 1988; Jara 1996, 2002.
4. Lévi-Strauss 1982.
5. Viveiros de Castro 1998; Bacigalupo 2016.
6. Crosby 1969.
7. Harper et al. 2011.
8. Washer 2010.
9. Wiener 2013; Latour 2010.

Bibliography

Bacigalupo, Ana Mariella. 2016. *Thunder shaman: Making history with Mapuche spirits in Chile and Patagonia*. Austin: University of Texas Press.

Basso, Keith H. 2009. Speaking with names: Language and landscape among the Western Apache. *Cultural Anthropology* 3 (2): 99–130.

Crosby, Alfred W., Jr. 1969. The early history of syphilis: A reappraisal. *American Anthropologist* 71 (2): 218–227. https://doi.org/10.2307/669854.

Harper, Kristin N., Molly K. Zuckerman, Megan L. Harper, John D. Kingston, and George J. Armelagos. 2011. The origin and antiquity of syphilis revisited: An appraisal of Old World pre-Columbian evidence for treponemal infection. *American Journal of Physical Anthropology* 146 (S53): 99–133. https://doi.org/10.1002/ajpa.21613.

Jara, Fabiola. 2002. The meaning of nominal animal categories among the Caribs of the Guianas. *Anthropos* 97 (1): 117–126. https://doi.org/10.2307/40465620.

Latour, Bruno. 2010. *On the modern cult of the factish gods*. Trans. Barbara Smith and E Weintraub. Durham: Duke University Press.

Lévi-Strauss, Claude. 1958. *Anthropologie structurale*. Paris: Plon.

———. 1982. *The way of the masks*. Trans. S. Modelski. London: Jonathan Cape.

———. 1988. *The jealous potter*. Trans. B. Chorier. Chicago: University of Chicago Press.

Lévi-Strauss, C. 1995. The structural study of myth. *New Studies in Aesthetics* 26: 85–96.

Malinowski, Bronislaw. 1948 [1926]. Myth in primitive psychology. In *Magic, science and religion, and other essays*, 93–143. London: Souvenir Press.

Santos-Granero, F. 1998. Writing history into the landscape: Space, myth, and ritual in contemporary Amazonia. *American Ethnologist* 25 (2): 128–148.

Viveiros de Castro, Eduardo. 1998. Cosmological deixis and Amerindian perspectivism. *Journal of the Royal Anthropological Institute* 4 (3): 469–488.

Washer, Peter. 2010. *Emerging infectious diseases and society*. New York: Palgrave Macmillan.

Wiener, Margaret J. 2013. Magic,(colonial) science and science studies. *Social Anthropology* 21 (4): 492–509.

9

French Malaise in the Taino Myths of Origin

We have examined Taino people's general understandings of zemes, and now we examine a myth of origin recorded by Pané in the Antilles between the second and the third Columbian travels (1495–1498), which describes specific zemes explicitly identified with syphilis and closely related treponemal diseases. This myth, which corresponds to the Pan-Amerindian legend of the Sun and the Moon, narrates the adventures of these zemes, contains references to the diseases they produced, the history of their treatments and interactions with humans, and their influence in the development of Caribbean societies. The identities of the zemes and the references are further emphasized when analyzed through Breton's dictionary.

In his search for the antiquities and idolatries of the Taino, Pané sought and found their myth of origin. The myth or myths he collected describe various events of the Taino cosmogony:

> where they came from, where the Sun and the Moon originated, and how the sea was made.[1]

C. E. Giraldo Herrera, *Microbes and Other Shamanic Beings*,
https://doi.org/10.1007/978-3-319-71318-2_9

The myth presents a high level of structural complexity, which Pané attributes to the lack of writing, preventing the Taino from telling things orderly, so that he feels he is writing first what should go later, and later what should go first. Thus, the myth compiled by Pané presents a difficult interpretation; nevertheless, it contains and connects many elements which become significant in the light of Breton, of other versions of the myth of the Sun and the Moon, the history of treponematoses, and our contemporary understanding of these diseases.

9.1 Where the Taino Came From

Pané starts elaborating on the myths, describing how most of the people in the island of Española (Haiti) came from a cave called Cacibajagua (The Stone of *Jagua*[2]) in the Mountain of Cauta.[3] The myth starts at a moment of distress:

> The people of Cauta are in the caves, hiding from the Sun. A guardian, called Mácocael, is keeping watch at night, but he takes too long to return, and the people of Cauta close the door on him. The Sun captures and transforms him into the *Genipa* Stone (*Cacibajagua*), which now stands at the entrance and gives the cave its name.

Then the people in the caves organize a series of excursions in which men go out to fish and to find *digo,* a bathing medicine.[4] However, the Sun captures these scouts and transforms them successively into *jobo* trees (*Spondías lutea*),[5] and into *yahubabayael,* a nightingale.[6]

The promoter of these excursions is Guahayona, a trickster anti-hero who is the brother-in-law of Anacacuya, the chief of *Cacibajagua*. Having fooled the men out of the way, Guahayona tricks the chief into looking for a *cobo*[7]—a giant sea-snail—and drowns him. Then, Guahayona convinces the women to leave with him and take only as much *güeyo*[8] leaves as they can carry. The women abandon their men and children. Calling for their mothers, the hungry kids transform into toad-like creatures. Guahayona leaves the women in the island of Matinino and continues alone on his trip home to Guanín.

9.2 Where the Sun and the Moon Came From

The following part of the myth focuses on the adventures of Guahayona. The versions of Pané's account in Anghieira and Ulloa vary slightly in this segment. In Anghieira's version,[9] by a miraculous grace, Guahayona remains unscathed and unchanged (unlike the others who turned into stone, trees, bird, or toads). He travels through diverse errands and, along his voyage, he finds a beautiful woman at the bottom of the sea. He dives to her and obtains from her marble stones, called *cibas,* and golden tin plates, called *guanines.* These were considered sacred collars.[10]

In Ulloa's version, on his trip back home,[11] Guahayona rejoices on meeting Guabonito, a beautiful woman he had left behind at sea. They enjoy themselves until he falls sick with what 'we call *French malaise*'[12] (i.e. syphilis). Guabonito places him on *guanara,* and Ulloa understands this to be in isolation. However, Arróm notes that in Guajiro, another Arawakan language, *guanöru* is disease. Moreover, Ruy Díaz de Isla in one of the earliest treatises on the what he called the serpentine disease (pox or syphilis) records *guaynara* as one of the names given to this disease. This would indicate that Guabonito placed the pox on Guahayona.[13] Whether Guabonito gave him the pox, or placed him in isolation, subsequently she cures him of his sores.

Having overcome the disease, he acquires a new name and becomes Albeborael Guahayona. Guabonito then requests his permission to continue her trip, which he grants. She then awards him with treasures to adorn his arms: *cibas* (*cobo* seashell slabs) and *guanines* (round, flattened, and perforated pieces of metal). These guanines originated from Guabonito, Albeborael Guahayona, and his father Hiauna.[14]

Interested in the metal but expressly confused by the mixed kinship relations, Pané transcribes literally what he is told about its origin:

> The son of Albeborael by the side of the father was called Híaguaili Guanín, meaning son of Hiauna, ever since the metal has thus been called.[15]

Breton's dictionary illuminates the myth in various ways. First, it contains a reference to a *Hiàli* (the French pronunciation corresponds to the Spanish *Híguaili,* the name of the metal), in a version of the myth of the

Sun and the Moon. In the myth recorded by Breton, the Moon visits a girl in dreams and makes her pregnant. However, the mother of the girl grows suspicious and one night leaves a guard who catches the couple. The Moon takes flight, but the guard stains his skin with *jagua, Genipa americana* juice, leaving the marks the Moon still bears. In shame, the Moon flees to the sky. From the incestuous relation is born *Hiàli,* the brilliant Sun, founder of the Carib nation, who is delivered to his father by a hummingbird.[16]

The myth of the Sun and the Moon figures widely in mythologies through the Americas.[17] Lévi-Strauss analyzed some of its multiple variations[18] before making an extensive study of them in the *Jealous Potter.*[19] The basic structure of the myth involves a love triangle or an otherwise transgressive relation, either of incestuous or extra-human character. The transgression leads to jealousy, to the blemishing of the characters, their flight, and their transformation into heavenly bodies with their distinguishing spots. This process brings about the origin of pottery or metallurgy and of agriculture or fishing and generates an infectious disease producing terrible rashes, which in recent versions remains unidentified.

The myth recorded by Pané contains all the characteristic elements of the myth. However, Pané explicitly associates Guahayona's affliction with the *French Malaise* (i.e. syphilis). He declares this disease was well known in the islands. The friar describes the disease as being similar to leprosy, producing rashes and sores, matching the European medical descriptions of the time. He points out, however, that it did not affect the natives as severely as it did Europeans.[20]

The myth makes various other allusions, which are consistent with syphilis or other treponemal diseases. The transformations of the children of Cauta into toads, animals with warty skin, describe succinctly the rashes of the secondary stage of treponemal diseases. The transformation of the men into rock, trees, and bird suggests the paralysis of the insane characteristic of the neurosyphilis in the tertiary stage of this disease.

Like Pané, Breton,[21] a century later, indicates that syphilis and yaws (another treponemal disease) were well known to the Callinago, who referred to them by the name of *yáya.* Nevertheless, these diseases did not affect the islanders as severely as they did Europeans, which he attributed to the weather and to the multiple treatments they had for them. The Callinago called the sores of the *yáya: Iaoüami,* and to have them: *Iaoüámitti,* which they regarded as a sign of laziness.[22] Breton explains that the Callinago protected themselves from the disease by blackening their skin with tinctures of *chaoua* (*Genipa americana).* They also massaged those suffering from the disease with the juice of the bark of *chipíou,* an extremely bitter tree. Finally, to avoid deformation, they applied the glue of marine snail shell and the bark of *mibi* lianas in a cotton plaster to close the open wounds. The use of *Genipa. a* and *Strombus* spp. seashells as antisyphilitic remedies[23] further corroborates the identification of the disease in the myth with treponematoses.

Breton and Rochefort also mention that the Callinago treasured a metal called *calloúcouli* or *caracoli* which:

> was sent to France to know and imitate, but it was in vain, our Naturals, always recognized the fraud. Silver, even the very gold, will not call their attention compared with this one. This metal has the advantage that neither oxide nor green attack it ever, nor achiote, nor trash dirty it, in such a way that with just wiping it with the hand it becomes clean. From it, they make half-moons they hang to their necks, and are the most expensive of their jewels. From it, they make straws of half inch thick they hang onto their perforated nose. A man that has a calloúcouli believes himself rich, with one calloúcouli a slave can be obtained from them, this product is uncommon and precious for them, any other is nothing in comparison with this.[24]

This ornament went together with a *Iámourourou:* collar made from the spine of a fish and elongated, transparent grains. The description of these adornments matches that of the *guanines* and the *cibas.*

9.3 How the Sea Was Made: The Caracaracol and "Yaya" Master of Manioc and the Sea"

The following part of the narrative turns back to Cauta and elaborates on how the island came to have women, how the sea was created, and on the adventures of Deminan Caracaracol, the cultural hero of the Taino:

> The men were eager for women, but they could not find any tracks of them. One day while bathing, they see some ungendered beings. The men try to catch them, but the beings run away like eels. Some men follow the track and determine that there are four beings. Four Caracaracol men are assigned the task of capturing them. Pané describes caracaracol, as a scabby disease that makes the skin rough and the hands horny. The horny hands allow Caracaracol men to capture the ungendered beings. Then *inriri,* a woodpecker,[25] carves vulvas into these beings, transforming them into women.

He proceeds to tell about the origin of the ocean:

> A man called Yaya,[26] who had no name, had a strife with his son Yayael. Yayael had tried to kill his father and for this was banished. After Yayael defies banishment, Yaya kills him and places his bones in a funerary calabash. After some time, Yaya and his wife realize that the bones have transformed into fish, which they eat.

Four brothers, born of a single birth that killed their mother, hear about the fish and plan to steal the calabash in the absence of Yaya. Déminan, who was born caracaracol and is the leader of the troop, performs the feat. The brothers eat the fish. But noting Yaya's return, they attempt to restore the calabash to its place. They fail, the calabash falls and cracks, leaking its containments: water and fish that fill the ocean.

As the four brothers flee from Yaya, they come across the door of Bassamanaco.[27] Seeing the old man is coming back with cassava,[28] they say, 'let us meet our grandfather'. Déminan follows the old man into his house and asks him for some manioc bread. But [B]Ayamauaco becomes offended by the request and instead throws him *guanguayo,* a spittle with *cohoba* wounding Déminan on the back. Déminan returns to his brothers and tells them what occurred with *Bayamanacoel.* Déminan's torso swells, threatening his life. The brothers decide to operate. They open the wound with a stone axe and extract from it a female turtle. They make a house and raise the turtle. In Ulloa's version, Pané does not know what happens afterwards. In d'Anghiera's version, it is a woman that is born out of the sore and 'all the brothers use her mutually, and from her engender sons and daughters'.

Desowitz suggests that the condition of *caracaracol* corresponds to pinta or carate.[29] This is another treponematosis which, although closely related with syphilis, constitutes a milder affliction that solely affects the skin, changing its pigmentation and producing callousness. This medical condition has been highly regarded by many Amerindian and Mestizo communities.

Oviedo presents the earliest reference to carate while describing the Cueva Indians of Panama; Cueva chiefs were transported in hammocks, by a pair of carriers:

> The Indians employed on this task are generally slaves or naborías, who are semi-slaves forced to serve; and the Indians who serve as hammock carriers, they seek them *carates*… *Carate* is called the Indian who naturally has his whole person or most of it pealed, and their skin flaking in the manner of *empeynes*. They appear ugly, but in general are hardy and with better strength, and they appear friezed, and this frieze, is a malady that ends when this itch or disease has gone through all their body and they have changed all their skin.[30]

In a similarly perplexed vein, Father Rivero stated of the Tunebos or U'wa of the Orinoco River in the Colombian eastern plains:

> In this region, they suffer from a filthy and repulsive disease called *carate*, and it is of the manner of leper, that they are covered to face and hands, with blue and white spots, that horrify vision. They are so savage in everything that they price themselves and make gallantries on such disease. To such a degree, that if any maiden in their town does not have *carate*, nobody wants her or sees her as a woman of good treatment, and in order for her not to lose marriage, they give her a drink out of which she gets *carate*. Then without further patrimony or dote than this she finds at her convenience and purposes so many pretenders as if she had in *carate* ownership of a mayoralty, a county or the Flemish states.[31]

Carate is also known as cara-cara, puru-puru, pinta, and various other local names; today it is endemic in Central America and Northern South America. Like the U'was or Tunebos of the nineteenth century, various communities in the Amazon regarded pinta as an ethnic identifier and intentionally fostered its contagion during puberty rituals.[32]

9.4 Guayanara, Yaya, and Caracaracol, Treponematoses and Their Treatments

Callinago language offers additional information to understand the Taino myth. The name of the Taino cultural hero Deminan Caracaracol resembles the *caracoli* or *calloúcouli,* the Callinago's most treasured adornment, replicating the relation between Taino metal *Híguaili guanin* and *Hiàli,* the Callinago cultural hero. In Callinago *yáya*—the Taino name of the father of the sea and owner of the cassava—was the name for syphilis and yaws.[33] The wound resembling a turtle-shell Yaya inflicts on Déminan, spitting sputter with cohoba on his back, evokes the initial sore of treponemal diseases, in particular of yaws. Known as mother-yaw, this initial lesion can be quite prominent; it is usually described as a raspberry, a mosaic shape which bears some resemblance to a turtle-shell.

From the transformations of the people of Cauta into stone, trees, bird, and toads, through the rashes affecting Guahayona after some pleasurable time with Guabonito, to the condition of the Caracaracol men and Déminan's wound, disease is a constant theme. Moreover, all these afflictions can be expressions of closely related diseases or even of a single disease. Albeit the descriptions are too vague to warrant definitive conclusions, there are various explicit allusions to treponemal disease and its treatments.

Cross-referencing the myths collected by Pané with Breton's dictionary corroborates and further emphasizes the role of treponemal disease in the Taino myths of origin (see Fig. 9.1). Yáya (syphilis or yaws) could be interpreted as the father or master of the sea. This identification is coherent with animic notions of masters of game, as pathogenic entities, who defend their protégées through disease. Fishing, shelling, and seafaring involve conditions which facilitate the transmission of yaws and other treponemal diseases.

The name Cacibajagua, 'The rock of the *jagua*', into which the guardian Mácocael is transformed by the Sun, or in Breton's version the guard who stains with *jagua* the face of the Moon, refers to a means of protection against the disease: the tincture of *jagua*.[34] Furthermore, the *cobo* shell, which the chief is looking for when Guahayona drowns him, and

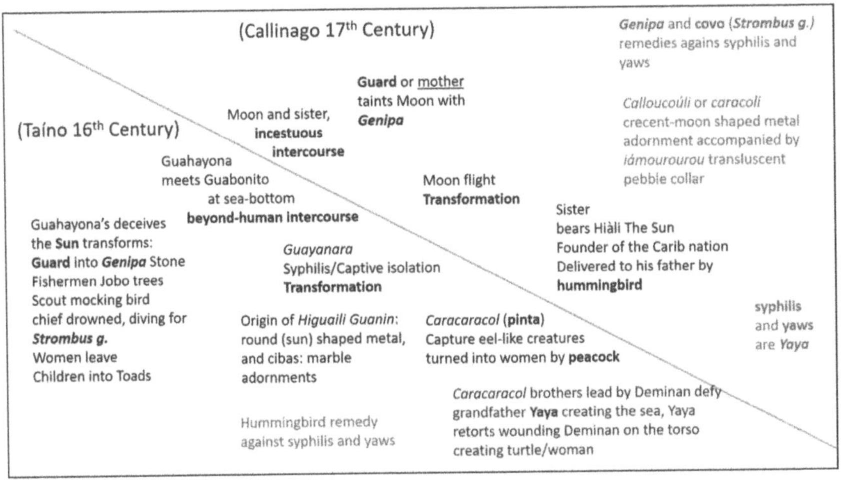

Fig. 9.1 Symmetries between the Taino and the Callinago myths of the Sun and the Moon

the *cibas* awarded to the latter by Guabonito, have a medical function healing the wounds of treponemal disease. This confirms the interpretation of the first part of the myth—where the Taino hide from the Sun—and describes Taino interactions with treponemal disease. Their transformation into stones, trees, birds, and toads depicts the effects of treponemal disease, and their strong concern for medicinal baths and purgatives attempts to remedy it.

Pané noted that the story does not seem to follow a chronological order. It is indeed possible that the myths were arranged in reverse chronological order, with new at the beginning and older stories narrated at the end. The story of how the islands came to have women would be a logical development of the story of Déminan and his brothers. The story of Cacibajagua could be a continuation of the story of Guahayona. On the other hand, Lévi-Strauss has shown how neighbouring versions of the same myth often present substantial differences in their narrative and in other structural features; he argues convincingly that such structural differences codify political relations and different perspectives on the same stories.[35] Alternatively, these paradoxes could also imply that the myths describe a dynamic rather than an origin. Various Amerindian groups

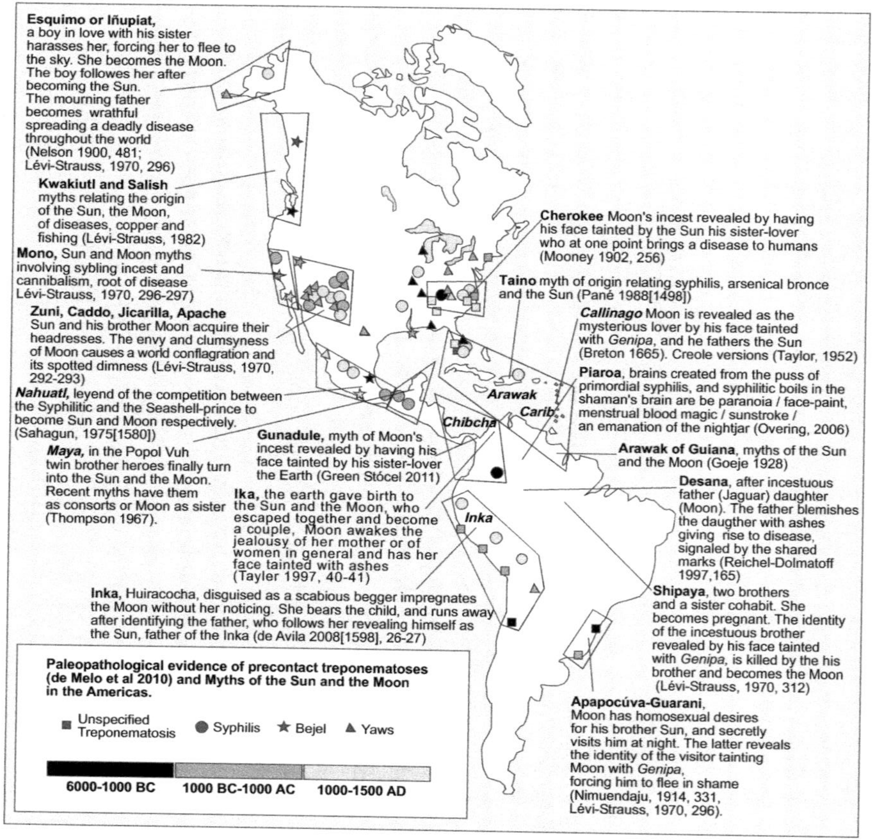

Fig. 9.2 Map of the distribution of paleopathological evidence of pre-contact treponematoses in the Americas and myths of the Sun and the Moon

claim that myths do not solely pertain to a remote past, but that they are rather a concurrent dynamic constantly re-enacted by shamanic beings at different dimensions of reality, with consequences in health, as well as in political and environmental relations.[36]

The analysis of paleopathological evidence in archaeological remains confirms the presence in pre-Contact Haiti of at least one kind of treponematosis. Treponematoses are a group of closely related diseases, including syphilis, produced by varieties of the bacteria *Treponema pallidum*. The remains found in the Antilles present the characteristic treponemal bone alterations affecting juveniles and adults. These alterations are otherwise coupled with signs of good health and of physical strength.[37] Thus, the archaeological remains confirm the diagnosis of the friars, both regarding the general nature of the disease and its mild effects on Taino and Callinago peoples. Given that the disease affected the bones, it could potentially be syphilis (*T.p.* subsp. *pallidum*), yaws (*T.p.* subsp. *pertenue)*, or bejel (*T.p.* sub. *endemicum)*.

The myth of the Sun and the Moon contains multiple allusions referring to syphilis and other treponematoses. Furthermore, the distribution of paleopathological evidence of pre-contact treponematoses through the Americas corresponds with the cultural areas where myths of the Sun and the Moon have been collected (See Fig. 9.2).

Notes

1. My translation from (Pané 1988b [1500], 38).
2. *Genipa americana* (Grandtner and Chevrette 2013, 262).
3. Taino is regarded as an Arawak language (Taylor 1954). *Kauta* is the Guyana Arawak name for *Licania ochtandra,* a small tree, and the ashes of the bark are employed together with clay to make pottery (Grandtner and Chevrette 2013, 353).
4. Pané 1988b [1500], 38. Perhaps Guyana Arawak *chigonet* (*Bellucia grossularioides*), which is antiseptic and fungicide (Grandtner and Chevrette 2013, 63).
5. Grandtner and Chevrette 2013, 625.
6. Presumably *Mimus poliglotos.*

7. Arróm (p. 74) identifies cobo as *Strombus gigas,* which is frequently employed in the making of what are assumed to be three-pointer stone zemes (Breukel 2013).
8. Arróm (p. 74) indicates the Indians of Guyana denominate *weya*, a riverside algae (*Mourera fluvialis,* Aubl.), the ashes of which are used to chew tobacco; however, this does not fit Pané's description. The text notes that güeyo leaves are similar to albahaca, and were used as purge and bathing medicines through a 7-day fasting course (Pané 1988a [1498], 37, 38, 48). Alternatively, we could suggest black *maho* (*Annona exsucca*), which the Arawak of Guyana denominate *hueyechi* or *koyechi* (Grandtner and Chevrette 2013, 37). The leaves of this plant display antiprotozoal and cytotoxic properties (Osorio et al. 2007); other plants of the same gender have been shown to have antibiotic properties (Takahashi et al. 2006).
9. Pané 1988b [1500], 91.
10. Pané 1988b [1500], 64.
11. The text is ambiguous regarding the place, whereas the title of the chapter states that he was travelling back to Cauta, while all the narrative before and after suggests he was travelling to Guanin.
12. Pané 1988b [1500], 25.
13. Pané 1988b [1500], 63; Díaz de Isla 1539, iii.
14. Pané 1988b [1500], 26.
15. My translation from (Pané 1988b [1500], 26).
16. Troiani 2001, 112. Goeje (1942, 1928) translates the Guyana Arawak Híali as: he-who-has-become-luminous. There is also a version of these myths which persists amongst Black Caribs of Dominica; in this version, the *Genipa* juice is replaced by sooth, applied by the mother of the girl, and the Moon is revealed to be her brother, explaining through the incest his embarrassed flight (Taylor 1952).
17. Nelson 1900; Nimuendaju 1914; Mooney 1902; Tayler 1997; Thompson 1967; Green Stócel 2011; Reichel-Dolmatoff 1997; de Avila 2008 [1598]; Overing 2006.
18. Levi-Strauss dedicates to this myth numerous passages (Lévi-Strauss 1970, 296–299, 1982).
19. Lévi-Strauss 1982, 1988.
20. Pané 1988b [1500].
21. Troiani 2001, 185.
22. Troiani 2001, 35.

23. *Genipa americana* is widely employed as an antisyphilitic (Grandtner and Chevrette 2013, 262). In the nineteenth century, von Martius registered that baths with decoctions of the fruit of *Genipa a.* were employed to cure the rashes and ulcers produced by syphilis, in the Brazilian Amazon (Breitbach et al. 2013).
24. My translation (Breton and Troiani 2001, 34–35).
25. *Melanerpes striatus.*
26. Referring to Goeje (1928), Árrom points out that Yaya might be related to the Guiana Arawak Ia, meaning spirit, or essence, and infers that the reduplication implies spirit of spirits, and so supreme being.
27. Arróm notes that Ulloa employs different spellings through the text (Bassamanaco, Aiamauaco, Baiamanicoel), but assumes this is an inconsistency of the translator and homogenizes spelling for the sake of clarity. However, considering that these different spellings are found in a single paragraph it must be considered that the story may involve different referents, or that the spellings may reflect distinct grammatical forms, as occurs in Callinago.
28. *Manihot esculenta* bread
29. (*Treponema p. sbsp carateum*) in Desowitz (1998).
30. Fernández de Oviedo 1959 [1535], T 2, 2 part, Book XXIX Chapter XXVI My translation.
31. Rivero 1883 [1736], 57 My translation.
32. Biocca 1945.
33. Respectivelly, *gross vérolle* and *pians* (Breton 1666, 291 & 399).
34. Breton and Troiani 2001, 64.
35. Lévi-Strauss 1982.
36. Viveiros de Castro 1998; Bacigalupo 2016.
37. Drew 2009.

Bibliography

Bacigalupo, Ana Mariella. 2016. *Thunder shaman: Making history with Mapuche spirits in Chile and Patagonia.* Austin: University of Texas Press.

Biocca, Ettore. 1945. Estudios etnobiológicos sobre os indios da regiao do alto rio Negro, Amazonas nota II: transmissao ritual e transmissao criminusa de espiroquetose discromica (puru-puru, Pinta, etc) entre os indios do rio Icana. *Separata de Arquivos de Biología* 29 (265).

Breitbach, Ulrike B., Michael Niehues, Norberto P. Lopes, Jair E.Q. Faria, and Maria G.L. Brandão. 2013. Amazonian Brazilian medicinal plants described by C.F.P. von Martius in the 19th century. *Journal of Ethnopharmacology* 147 (1): 180–189. https://doi.org/10.1016/j.jep.2013.02.030.

Breton, Raymond. 1666. *Dictionaire français-caraïbe*. Auxerre: Gilles Bovqvet, Imprimeur ordinaire du Roy. Original edition, 1666.

Breton, Raymond, and Duna Troiani. 2001. *El caribe insular del siglo XVII Tratado sobre la lengua y la cultura de los Callínago Traducción al español del Dictionnaire caraïbe-français de Raymond Breton*. París: CELIA-CNRS. Original edition, 1665.

Breukel, Tom. 2013. Threepointers on trial: A biographical study of Amerindian ritual artefacts from the pre-Columbian Caribbean. Master's thesis, Archaeology, University of Leiden.

de Avila, Francisco. 2008 [1598]. *Ritos y tradiciones de Huarochirí: manuscrito quechua de comienzos del siglo XVII: versión paleográfica, interpretación fonológica y traducción al castellano*, vol. 12. Lima: Instituto de Estudios Peruanos: Instituto Francés de Estudios Andinos.

Desowitz, R.S. 1998. *Who gave Pinta to the Santa Maria? Tracking the devastating spread of lethal tropical diseases into America*. London: Harcourt Brace.

Díaz de Isla, Ruy. 1539. *Tractado co [n] tra el mal serpentino que vulgarmente en España es llamado bubas*. Lisboa: (H)ospital de Todos los Santos.

Drew, Rose. 2009. From Saladoid to Taino: Human behavior from human remains in the Greater Antilles. *Bulletin of the Peabody Museum of Natural History* 50 (1): 167–185.

Fernández de Oviedo, Gonzalo. 1959 [1535]. Historia general y natural de Indias. In *Biblioteca de Autores Españoles*. Madrid: Biblioteca de Autores Españoles. Original edition, 1935.

Goeje, Claudius Henricus. 1928. *The Arawak language of Guiana*.

———. 1942. De Inwijding tot Medicijnman bij de Arawakken (Guyana) in Tekst en Mythe. *Bijdragen tot de Taal-, Land- en Volkenkunde* 101 (1): 211–276.

Grandtner, M.M., and J. Chevrette. 2013. *Dictionary of trees, volume 2: South America: Nomenclature, taxonomy and ecology*. New York: Elsevier Science.

Green Stócel, Abadio. 2011. *Anmal Gaya Burba: Significados de vida*. Ph.D., Educación, Universidad de Antioquia.

Lévi-Strauss, Claude. 1970. *The raw and the cooked*. Trans. J. & D. Weightman. London: Harper & Row. Original edition, Le cru et le cuit.

———. 1982. *The way of the masks*. Trans. S. Modelski. London: Jonathan Cape.

———. 1988. *The jealous potter*. Trans. B. Chorier. Chicago: University of Chicago Press.

Mooney, James. 1902. *Myths of the Cherokee*, vol. 19. Washington, DC: Smithsonian Institution. Bureau of American Ethnology.

Nelson, Edward William. 1900. *The Eskimo About Bering Strait*. Washington, DC: U.S.G.P.O.

Nimuendaju, Curt. 1914. Die Sagen von der Erschaffung und Vernichtung der Welt als Grundlagen der Religion der Apapocúva-Guaraní. *Zeitschrift für Ethnologie* 46 (2/3): 284–403.

Osorio, Edison, Gabriel Jaime Arango, Nora Jiménez, Fernando Alzate, Grace Ruiz, David Gutiérrez, Marco Antonio Paco, Alberto Giménez, and Sara Robledo. 2007. Antiprotozoal and cytotoxic activities in vitro of Colombian Annonaceae. *Journal of Ethnopharmacology* 111 (3): 630–635.

Overing, Joanna. 2006. The stench of death and the aromas of life: The poetics of ways of knowing and sensory process among Piaroa of the Orinoco Basin. *Tipití: Journal of the Society for the Anthropology of Lowland South America* 4 (1): 2.

Pané, Ramón Fray. 1988a [1498]. Relación de Fray Ramón acerca de las antigüedades de los indios, las cuales, con diligencia, como hombre que sabe la lengua de ellos, las ha recogido por mandato del Almirante. In *Relación acerca de las antigüedades de los indios: El primer tratado escrito en América*, ed. Jose Juan Arrom. Bógota: Siglo XXI Ediciones.

———. 1988b [1500]. Relación de Fray Ramón… In *Relación acerca de las antigüedades de los indios: El primer tratado escrito en América*, ed. Jose Juan Arrom. Bógota: Siglo XXI Ediciones.

Reichel-Dolmatoff, G. 1997. *Rainforest shamans: Essays on the tukano indians of the Northwest Amazon*. Cambridge: Green Books.

Rivero, Juan de. 1883 [1736]. *Historia de las misiones de los Llanos de Casanare y los ríos Orinoco y Meta*. Bogota: Biblioteca de la Presidencia de Colombia. http://www.lablaa.org/blaavirtual/historia/hismi/indice.htm.

Takahashi, Jacqueline A., Cássia R. Pereira, Lúcia P.S. Pimenta, Maria Amélia D. Boaventura, and Luiz G.F.E. Silva. 2006. Antibacterial activity of eight Brazilian annonaceae plants. *Natural Product Research* 20 (1): 21–26. https://doi.org/10.1080/14786410412331280087.

Tayler, Donald. 1997. *The coming of the sun: A prologue to Ika sacred narrative*. Oxford: Oxford University: Pitt Rivers Museum.

Taylor, Douglas. 1952. Tales and legends of the Dominica Caribs. *The Journal of American Folklore* 65 (257): 267–279.

———. 1954. A note on the Arawakan affiliation of Taino. *International Journal of American Linguistics* 20 (2): 152–154.

Thompson, J. Eric S. 1967. Creation myths (part 2). *Estudios de cultura maya* 6: 15–43.

Troiani, Duna. 2001. *El caribe insular del siglo XVII Tratado sobre la lengua y la cultura de los Callínago. Traducción al español del Dictionaire caraïbe-français de Raymond Breton*. Trans. Duna Troiani. París: CELIA-CNRS Original edition, 1665.

Viveiros de Castro, Eduardo. 1998. Cosmological deixis and Amerindian perspectivism. *Journal of the Royal Anthropological Institute* 4 (3): 469–488.

10

The Spotted Sun and the Blemished Moon, Nahuatl Views on Treponematoses

This chapter explores Nahuatl understandings of *nanahuatl*, syphilis, and other treponemal diseases, which were understood as the tutelary deity of the Nahuatl peoples. It examines their classification and association with sectors of Mexican society, their treatments, and how this knowledge was codified in the myth of the Sun and the Moon. This version of the myth can also be interpreted as a description of the transition between different technologies employed to make precious objects out of metal, a process leading to the development of arsenical bronze, a material was employed in the treatment of treponemal diseases.

A clearer and more detailed view on pre-Contact Amerindian relations with treponematoses can be found in various Mexican codices, in particular in the Florentine Codex. This document is the result of one of the earliest and most ambitious ethnographic projects: over the course of more than 40 years, Bernardino de Sahagun directed a group of Amerind scholars and artists in the compilation, translation, and illustration of Nahuatl knowledge. In this document, the pox features various times. The codex indicates that the Nahuatl knew the pox as *Nanáhuatl*. *Nanáhuatl* was the Sun and regent of the Nahuatl. The codex Yoalli Ehēcatl/Borgia presents an illustration of the syphilitic Sun nourishing the different Nahuatl gods

C. E. Giraldo Herrera, *Microbes and Other Shamanic Beings*,
https://doi.org/10.1007/978-3-319-71318-2_10

Fig. 10.1 Nanahuatl the syphilitic Sun drawn from *The Book of Night and Wind* (Codex Yoalli Ehēcatl/Borgia, 45) and its facsimile (1825–1831, 43)

and the corresponding segments of Nahuatl society (see Fig. 10.1). *Nani* is the mother, master, or regent. The term *Nahuatl* had the meaning of tone, of good sound, and by extension brilliance and meaning – it is the word through which the Nahuatl refer to themselves. Nahuatl physicians distinguished three varieties of the disease[1]:

(a) The noble or gentle, *tekpil-nanáhuatl,* did not cause pains or facial deformations.
(b) The common or dirty, *tlacazol-nanáhuatl,* caused deep open wounds.

(c) The cruel or tradesmen's, *pocho-nanáhuatl*, produced profound lesions and deep pains, crippled hands and feet, and resided in the bones.

This classification relates particular pathologies with specific segments of Nahuatl society, indicating different epidemiologic contexts. *Tekpili was* the basic nobility, literally the offspring of noble gentry; however, the term also designates the condition of being well-reared.[2] *Tlacazol* is dirt and excrement, *tlaçotla* is love, and *tlacotli* is slave. The Nahuatl associated dirt with lust and creativity. *Tlazoltéotl,* the dirt-eater, is described as a lunar or tectonic deity of lust, excrement, maternity, pottery, weaving, and prostitution.[3] Finally, the *pochtecatl* were merchants who, in whole family units, carried out long-distance trade in large Ceiba canoes.[4] This classification describes the characteristics of treponematoses affecting the region.

As was pointed in the last section, regarding *caracaracol* and *carate,* until the twentieth century, pinta was considered an honoured and desirable condition by many Amerindian and mestizo communities in Central and South America. Carrillo points out that during the Mexican revolution, *pintado* battalions, conformed by men with pinta, were very highly regarded.[5] Osteopathological evidence, on the other hand, indicates the presence in the Mexico Valley of the three other treponematoses. Bejel was a common disease in the Mexico Valley and the Marismas Pacific Coast for two millennia, from 1100 BC to 1150 AD. It disappears thereafter, replaced by yaws, which dominates the panorama for about 150 years. By about 1300 AD, yaws starts sharing space with syphilis. This trend continues after the Encounter at the time when the codex was written.[6] Seafarers, like the pochtecatl, often remain with their extremities exposed, humid and at low temperatures, which are conditions that facilitate the transmission of yaws, a disease that affects the bones, crippling hands and feet. On the other hand, *tlacazol,* which associates the disease with love, lust, and prostitution, strongly suggests syphilis. The association of the disease with slavery may describe the way the Nahuatl and other Amerindian communities grew by enslaving or incorporating other peoples, thus suggesting that the disease played a role in imperialism as is also insinuated by Pané's myth.[7]

Besides their classification of the disease, the Nahuatl also had various means of treating *nanáhuatl,* specifically denominated *nanáhuapatli*. These medicines were rapidly collected and highly valued in Europe.

Hernandez, physician to Philip the II, King of Spain, registers at least eight herbal and floral *Nanáhuapatli.*[8] The Florentine Codex, one of Hernandez's main sources, mentions at least six of them:

> The disease is cured drinking the water of *tletlemaitl* herbs, and taking baths with either dust of *tlaquequetzal* or *scrapings of copper*... when the rash brakes out the diseased should drink *atolle* with a seed denominated *michivauchtli* or alternatively drink water of *quauhtlepatli* roots, five times a day as well as baths... If the diseased becomes paralyzed, he should drink water of the *tlatlapanatic* root, and undergo bleeding thereafter.[9]

This extended arsenal reflects a continuous experimentation, but it is the myth that offers an insight into how Nahuatl understood the disease, its origins, and developments.

Nanáhuatl or Nanahuatzin is a syphilitic deity, the humble orphan of Quetzalcoatl, the god of wind and storms, adoptive son of Xochiquétzal and Piltzintecuhtli. Xochiquétzal, whose name translates flower-feather, was a deity of beauty, eroticism, liaisons, and the crafts. She was patroness of weavers, painters, woodcarvers, and smiths,[10] arguably also of prostitutes as she was also referred to as Tlazoltéotl.

Xochiquétzal required devoted labour, and produced the rashes of the incurable pox on the weavers who broke the fasting they owed to her.[11] Piltzintecuhtli (hummingbird-prince of the flowers) was the male god of beauty and sexuality, of games, dance, singing, healing, and hallucinogenic drugs. Both deities are associated with cures against disease, like *izquixóchitl* (the cornflower, *Bourreira huanita Hemsl),* which is Xochiquétzal's most representative flower and part of the fasting her devotees had to follow.[12] Xochiquétzal and Piltzintecuhtli, beauty and its production, engender knowledge. Together, these deities of the crafts and arts represent the generation of beauty in ecological interactions involving interspecies communication and pollination, the relation between moving forms and shaping movements. The couple engenders agriculture, fathering Centeotl, the corn god. This suggests that the origin of agriculture was associated with the rise of contagious disease, which is consistent with recent theories about the evolution of disease and society.[13] Or, alternatively, that *nanáhuatl* was a manipulated disease.

Further indications of the domestication of the disease appear in the codex Magliabechiano, which describes the origin of Nanáhuatl: a bat, which came out of Quetzalcoatl's semen, was commissioned to bite the vulva of Xochiquétzal and cut a piece of her flesh. This flesh was brought to the gods who cleansed it. From the cleansing water came roses that did not smell good. Then the bat himself took a rose to *Mictlantecuhtli,* the god of death, who washed it and from its water came sweet-smelling roses, called *suchiles.*[14] Piltzintecuhtli could indicate an association between the effects of hallucinogenic drugs and those of the neuropathogen, suggesting a psychotropic disease.

Another version features in the codex Yoalli Ehēcatl/Borgia (Fig. 10.2). The codex portrays *Chalchíuhtlícue,* goddess of the flowing waters and merchants, carrying on her back a *tlacaxxotl* ceramic wind-oven, with a snake (representing a plague) protruding from the hearth. The goddess quarters, bleeds, and guts the snake, lighting fire on an incensory, which contains the symbol of the Sun. From the pieces of the snake grows a plant-like being. On top of the sacrificial table next to the plant sits naked *Xochiquetzal,* the goddess of flowers, love, sexuality, and crafts, and in the middle *Nanahuatzin,* the syphilitic Sun god, appears as a child, protruding from the top of the burning wind-oven.[15] The image could make

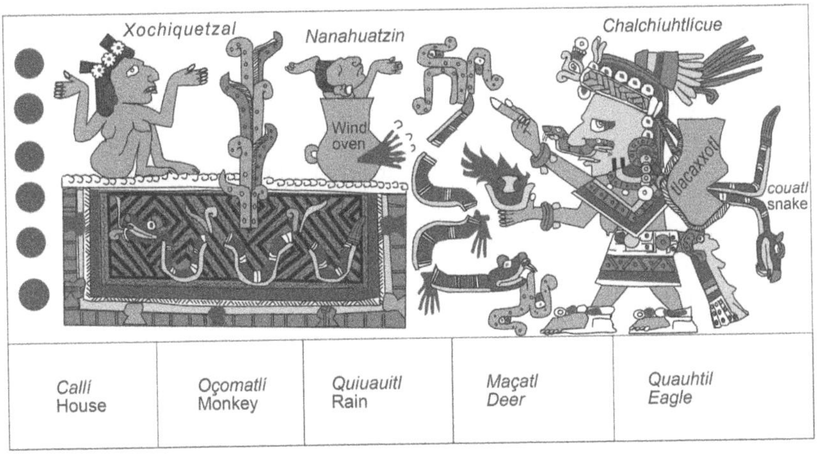

Fig. 10.2 Xochiquetzal, *Nanahuatzin,* and Chalchiuhtlicue (goddess of the waters) (Codex Yoalli Ehēcatl/Borgia, 23) and its facsmile (1825–1831, 19)

allusion to the acquisition of treponemal diseases and metallurgy through trade, and their subsequent developments into three distinct varieties.

Like Hiàli, Nanáhuatl or Nanahuatzin (the syphilitic lord) became a cultural hero, the Sun of movement.[16] This myth describes the transition of power at the end of the Toltec period and the start of the Aztec period.[17] I will summarize his myth briefly:

> After the end of the fourth era, the world was again in darkness and the gods decided someone had to take the responsibility for illuminating it, sacrifice to the fire, and become the sun. Tecuciztecatl (meaning the one coming from the place of the marine shell) who was young, rich, and beautiful, eagerly volunteered. But another volunteer was needed. Someone suggests Nanahuatzin, an orphan and a humble and syphilitic god. Other gods backed the motion and ordered him to "volunteer." He accepts the order as a compliment. Both gods present offerings of penitence to feed the fire. Tecuciztecatl presents treasures, playing up the traditional offerings: quetzal feathers instead of fir branches; instead of balls of grass, golden ones; jade instead of maguey spines; red coral instead of blood-tainted things; and excellent *copal* (incense) instead of the common one. Meanwhile, the offerings of the poor Nanahuatzin downplayed the traditional offerings: branches of cheap cane instead of fir, balls not even of grass but of pine-spikes, agave spines that were only agave spines, and things tainted with his own blood. Finally, for incense he scratched his own wounds. Tecuciztecatl was to try first, but he cowered with the fire in the heat of the moment. Nanahuatzin jumped without hesitation into the flames and the fire consumed him. Then, Tecuciztecatl took courage, but he only reached the ashes. The gods waited to see from what direction the sun would come out, until Nanahuatzin appeared transformed in blinding warmth. Soon, however, Tecuciztecatl appeared transformed into a second sun. Under their combined light gods and humans were alike, which the gods did not approve of. One god slapped Tecuciztecatl in the face with a rabbit, transforming him into the dimmer moon with the rabbit marked on its face.

The Nahualt legend of the Sun and the Moon can be partly interpreted as a myth about metallurgy. Precious metals like gold (*teocuitlatl,* literally 'god's faeces', from *teotl* gods *cuitlatl* faeces) were strongly associated with the Sun and the Moon.[18] Around 1200 AD, near the time when the events narrated in this story presumably occurred, an important transition was taking place in West Mexican metallurgy. Until that time, despite being a metal-rich region, metal artefacts had been imported from South America. Mesoamerican smiths reworked old artefacts in cold, or made new objects, usually of an alloy of copper and gold, melting old ones and recasting them with lost wax techniques. But by about 1200 AD, Mesoamerican smiths started producing metal alloys from raw local minerals imitating techniques formerly developed in the Central Andes (Peru), casting arsenical bronze out of copper, arsenic, and alloys of copper and silver to imitate gold and produce a rich variety of silver-like effects. This allowed them to reproduce South American designs with more precision, replicating the finer and longer sewing needles, the more sophisticated bell designs, larger and more elaborate tweezers, and thinner and harder axes.[19] There was also a transition from a former emphasis in the exploration of the acoustic properties of these objects, towards experimentation with colour and brightness.[20]

Nanahuatzin and Tecuciztecatl could stand for two different ways of acquiring and working metals. Tecuciztecatl brought extremely ostentatious, although not very flammable, offerings. He was feeding the fire with the symbolic brightness of acquired wealth, imported artefacts, and reworked metal. On the other hand, brute copper, like Nanahuatzin, has a rough crust-like appearance. Nanahuatzin's offers were apparently poor but implied a real sacrifice and raised the fire required by the novel smithy techniques. Copper melts at higher temperatures than arsenical bronze and gold, at lower temperatures than either copper or arsenical bronze (Cu>Cu-Ac>Au). The temperature rose so much that Tecuciztecatl, wearing a precious golden tiara, bracelets, and necklace, did not dare throw himself into the fire. The rough Nanahuatzin, wrapped in paper and presumably delusional, did not hesitate. Only when the fire eased did Tecuciztecatl throw himself in. Melting pure gold after having melted mineral copper alloys would dim its brightness and stain it like the moon or Tecuciztecatl's face.

This story of metallurgy is intimately related with treponematoses, which it mentions in various ways: from the syphilitic Nanahuatzin to Tecuciztecatl, who ends up bearing treponemal marks on his face. The myth also mentions various cures like the seashells to which

Tecuciztecatl's name refers. Seashells like *Spondylus* spp., perhaps the 'red coral', were highly luxurious items traded over long-distance networks.[21] They were considered 'the food of the gods'.[22] But the strongest link with treponematoses are metals themselves. As is noted above, the Florentine Codex mentions that the Nahuatl cured *nanáhuatl* (treponematoses) with baths of copper scrapings.[23] Mesoamerican artefacts were characterized by their extremely high arsenic content. While in the Central Andes arsenic content ranged only from 2% to 7%,[24] in Mesoamerican artefacts it reached up to 20%, crossing the thresholds of intoxication. Arsenic interferes with aerobic respiration producing hyperpigmentation and neuropathies,[25] akin to those produced by the American varieties of *Treponema pallidum*. Arsenic also interferes with the metabolism and reproduction of the pathogen.[26] Needles, pincers, and axes with high arsenic content were the craftspeople tools, bells those of dancers and musicians. It is thus not strange that their consecrated work with the excrements of the gods was associated with the diseases and simultaneously prevented their more nefarious expressions. Rather than two separate ways of dealing with disease, symbolic meaning and physical function were intrinsically related.[27]

It is tempting to follow Descola's classification of Nahuatl ontologies as a form of analogism, in which different realms of existence are connected by analogical relations.[28] The three different diseases and their correlations with particular sectors of society certainly seem to point this way. Also, highly suggestive is the explicit association of certain gods and goddesses with specific offices and crafts, or the identification of certain gods, diseases, or conditions, with particular ways to work and produce metal.

Although Mesoamerican and Andean ontologies differ from Amazonian ontologies, it could be suggested that the imperialist centralization of these societies may have made them prone to draw categorizations and equivalences. However, our cases also problematize Descola's assumptions. At no point in the aforementioned cases do the documents explicitly state the relations in terms of equivalence, such as *a* is to *b* what *x* is to *y*.[29] Characterizing the relations as forms of analogy would imply dismissing their inherent connections. It implies neglecting how they were understood, and maintaining the assumptions of Christian Western ontologies, for which gentleness or cruelty are exclusively human qualities which can only be applied to non-humans metaphorically or analogically. Classifying an ontology as analogism can be a form of negating its reality.

Furthermore, even from the point of view of natural sciences, the association of certain treponemal conditions or diseases with particular sectors of society would not be a case of analogy but of synecdoche. That is, the Nahuatl were not interrelating unrelated classifications of distinct kinds of phenomena; they were showing how each of the diseases corresponded to the sectors of the society which these diseases were likely to affect. Each of the diseases affected the sectors of society in question: *tekpil*-pinta was not just a gentler condition but was a cultivated condition of the nobility; *tlacalzol*-syphilis was related with love and it was venereal; and *pocho*-yaws was not just cruel, like a trader, but also it was the disease associated with maritime trading. In a similar manner, the association of certain gods and goddesses with particular trades is given by the means of treatment embedded in these offices.

Notes

1. Sahagun 1975 [1580], 1577, 221–225 Vol. X, Chapt. 28, Paragraph 5, Fo. 109–111.
2. Carrasco 1976, 194.
3. Rincón and Peñafiel 1885 [1595], 41–43, 69, 72 & 88; Mentz 2009; Carrasco 1976, 199–200.
4. Carrasco 1976, 230–235; Thompson 1949; Sámano 1985.
5. Carrillo 2013.
6. Mansilla et al. 2000.
7. A parallel with the role of virus affections in European colonialism can be suggested, see Crosby (2004).
8. Elferink 2008.
9. Sahagun 1975 [1580], 221–25. Vol. X, Chapt. 28, Paragraph 5, Fo. 109–111.
10. Díaz Cíntora 1990, 14, 23 & 79.
11. Sahagun 1975 [1580].
12. Díaz Cíntora 1990, 14 & 22.
13. Harper and Armelagos 2013.

14. Rossell and Díaz 2003, 128; Díaz Cíntora 1990, 17.
15. Seler 1904, 287–288.
16. Garibay 1961; Sahagun 1975 [1580], 7:3–7; Kohler 1997; Lupo 1991.
17. Velazco 1998.
18. Hosler 1995.
19. Hosler 1999.
20. Hosler 1995.
21. Sámano 1985; Marcos and Norton 1982.
22. Paulsen 1974; Murra 2002.
23. Sahagun 1577, Vol. X 109.
24. Ziobrowski et al. 1996.
25. Idrovo 2005.
26. Hughes et al. 2011; Ehrlich 1913.
27. Descola 2013, 5.
28. Descola 2013, 208–213.
29. Such relations only appear when dealing with deities with different origins such as Tlazoltéotl and Xochiquétzal.

Bibliography

Carrasco, P. 1976. La sociedad mexicana antes de la conquista. In *Historia general de México*, ed. D.C. Villegas, 165–288. Ciudad de México D.F.: Colegio de Mexico.

Carrillo, Ana María. 2013. From badge of pride to cause of stigma: Combatting mal del pinto in Mexico. *Endeavour* 37 (1): 13–20. https://doi.org/10.1016/j.endeavour.2012.11.005.

Crosby, Alfred W. 2004. *Ecological imperialism: The biological expansion of Europe, 900–1900*. Cambridge: Cambridge University Press.

Descola, Philippe. 2013. *Beyond nature and culture*. Trans. J. Lloyd. Chicago: University of Chicago Press.

Díaz Cíntora, Salvador. 1990. *Xochiquetzal, Seminario de Estudios Prehispánicos para la Descolonización de México*. Ciudad de Mexico D.F.: Universidad Nacional Autónoma de México.

Ehrlich, Paul. 1913. Address in pathology, on chemiotherapy. *The British Medical Journal* 2 (2746): 353–359.

Elferink, Jan G.R. 2008. Ethnobotany of the Aztecs. In *Encyclopaedia of the history of science, technology, and medicine in non-western cultures*, ed. Helaine Selin, 826–833. New York: Springer.

Garibay, Angel María. 1961. *Llave del náhuatl: colección de trozos clásicos, con gramática y vocabulario, para utilidad de los principiantes*. Porrua: Ciudad de México D.F. http://mexica.ohui.net/textos/2/.

Harper, Kristin N., and George J. Armelagos. 2013. Genomics, the origins of agriculture, and our changing microbe-scape: Time to revisit some old tales and tell some new ones. *American Journal of Physical Anthropology* 152 (S57): 135–152.

Hosler, Dorothy. 1995. Sound, color and meaning in the metallurgy of ancient West Mexico. *World Archaeology* 27 (1): 100–115.

———. 1999. Recent insights into the metallurgical technologies of ancient Mesoamerica. *JOM Journal of the Minerals, Metals and Materials Society* 51 (5): 11–14.

Hughes, Michael F., Barbara D. Beck, Chen Yu, Ari S. Lewis, and David J. Thomas. 2011. Arsenic exposure and toxicology: A historical perspective. *Toxicological Sciences* 123 (2): 305–332. https://doi.org/10.1093/toxsci/kfr184.

Idrovo, A.J. 2005. Possible health effects associated with pre-Columbian metallurgy. *Biomédica* 25: 295–303.

Kohler, U. 1997. El fuego entre los aztecas. In *El fuego. Mitos, ritos y realidades: coloquio internacional Granada , 1–3 de febrero de 1995*, ed. José Antonio González Alcantud and María Jesús Buxó Rey, 199. Barcelona: Anthropos.

Lupo, A. 1991. El sol en Jerusalén. *La Palabra y el Hombre* 80: 185–196.

Mansilla, J., B.M. Rothschild, C. Pijoa, and C. Rothschild. 2000. Transitions among treponematoses in ancient Mexico. *Chungará (Arica)* 32: 167–174.

Marcos, J., and P. Norton. 1982. From the Yungas of Chinchaysuyo to Cuzco: The role of La Plata Island in Spondylus trade. In *Social and economic organization of the prehispanic Andes*, ed. D.L. Browman, R.L. Burger, and M.A. Rivera, 7–20. Oxford: B.A.R.

Mentz, B. V. 2009. Esclavitud y semiesclavitud en el México Antiguo y en la Nueva España (con énfasis en el siglo XVI). *Studia Historica. Historia Antigua* 25 (0).

Murra, John V. 2002. *El mundo andino: población, medio ambiente y economía*, vol. 24. Fondo Editorial PUCP.

Paulsen, Allison C. 1974. The thorny oyster and the voice of God: Spondylus and Strombus in Andean prehistory. *American Antiquity* 39: 597–607.

Rincón, Antonio, and Antonio Peñafiel. 1885 [1595]. *Arte mexicana*. México Oficina tip. de la Secretaría de fomento. http://ia331427.us.archive.org/2/items/artemexicana00rincrich/artemexicana00rincrich.pdf.

Rossell, C., and M.Á.O. Díaz. 2003. *Las mujeres y sus diosas en los códices prehispánicos de Oaxaca*. Mexico: Centro de Investigaciones y Estudios Superiores en Antropología Social (CIESAS) & Porrúa.

Sahagun, Bernardo. 1577. *Historia Universal de las Cosas de la Nueva España: Códice Florentino*. Florence: Medicea Laurenciana Library, World Digital Library. http://www.wdl.org/en/item/10096/.

———. 1975 [1580]. *Florentine codex: General history of the things of New Spain*. Santa Fe: School of American Research and University of Utah.

Sámano, Juan de. 1985. Relación Sámano. In *Verdadera relación de la conquista del Perú. Edición de Concepción Bravo*, ed. Concepción Bravo. Madrid: Historia 16. Original edition, 1527.

Seler, Eduard. 1904. *Codex Borgia, eine altmexikanische Bilderschrift der Bibliothek der Congregatio de Propaganda Fide*. 3. Berlin: Loubat.

Thompson, J. Eric S. 1949. Canoes and navigation of the Maya and their neighbours. *The Journal of the Royal Anthopological Institute of Great Britain and Ireland* 79 (1/2): 69–78.

Velazco, Salvador. 1998. La imaginacion historiografica de Fernando de Alva Ixtlilxochitl: Etnicidades emergentes y espacios de enunciacion. *Colonial Latin American Review* 7 (1): 33–58. https://doi.org/10.1080/10609169885007.

Ziobrowski, C., E. Cabanillas, and L.G. Palacios. 1996. Estudio de aleaciones cobre-arsénico. *Boletín del Museo del Oro* 41: 131–143.

11

The West, Syphilis, and the Other Treponematoses

This chapter examines how Europeans came to share the Amerindian realities, like their treponematoses with the biosocial dynamics these diseases fostered, and how these diseases together with Amerindian knowledge influenced European understandings of disease, working as an ontological scaffold for the development of microbiology. It starts in the sixteenth century, examining the syphilis pandemic, its relation with the early proposal of the *Theory of Contagion*, showing how European physicians borrowed and adapted Amerindian knowledge to the paradigm of humoral medicine. Then, we explore the rise of the syphiloids between the seventeenth and twentieth century, showing how European physicians challenged earlier paradigms developing the *Ontological Theory of Disease*. It concludes with the formulation of *Germ Theory* during the twentieth century and the identification of *Treponema pallidum* as the causal agent of treponemal diseases, conducing to the development of understandings which parallel sixteenth-century Amerindian understandings of *Nanahuatl*.

Generally venereal, chronic if untreated, frequently causing deformities and affecting mental faculties, syphilis is a terrifying and stigmatizing disease.[1] It brought havoc as an epidemic that violently scourged Europe

C. E. Giraldo Herrera, *Microbes and Other Shamanic Beings*,
https://doi.org/10.1007/978-3-319-71318-2_11

and Asia after the siege of Naples in 1495. This battle drew multitudinous cohorts of multinational mercenaries, facilitating the rapid dispersion of the disease across the globe.[2] By 1497, on account of the venereal plague (*lues venerea*), warnings and cautions towards prostitutes were being deployed across Europe, even in small peripheral communities like Aberdeen in the north of Scotland.[3] At the same time, the epidemic reached India. A few years later, syphilis reached China and brought Venice to a military and civil emergency.[4] Apparently, it affected all levels and degrees of society, influencing even fashion.[5]

Unlike other epidemics, syphilis did not vanish after its diaspora—it remained and chronically affected a sizeable proportion of Western urban populations even as they flourished economically, artistically, and intellectually throughout the Renaissance, the Enlightenment, and Modernity. Multiple important figures in the Western arts and philosophy during this period were known or strongly suspected of having suffered from syphilis through some of their most prolific years,[6] with the disease presumably influencing not only their personal work, but that of their artistic and philosophic circles as well.[7] The disease also had a deep impact on medical and scientific knowledge in Europe: it influenced Fracastoro's formulation of the theory of contagion, Sydenham's ontological theory of disease, and Ehrlich's success experimenting with organic arsenic predecessors of antibiotics.[8] Thus, despite being terrible, and despite the moral taboos associated with syphilis, civilization was equated with syphilization,[9] and *Treponema pallidum* subsp. *pallidum* (its causal agent) was inadvertently acknowledged as playing an important role in the development of modern Western society.[10]

Yaws (*T.p. pertenue*) and bejel (*T.p. endemicum*), two equally nefarious non-venereal relatives of syphilis, were recognized around the seventeenth century and terrorized poor rural European communities until the nineteenth century. They, then, disappeared from most of Europe and remained relegated respectively to the humid lands of the tropical belt and to desert regions of Africa and Asia. Pinta (*T.p. carateum*), a milder skin infection, has remained endemic in Central and South America. While its means of transmission remain mysterious, it has been highly regarded and is purposefully transmitted in many Amerindian communities, as it may provide immunity against more damaging treponemal diseases.

11.1 Sixteenth Century, Syphilis, and the *Theory of Contagion*

Mutual accusations featured in the history of syphilis, as evidenced by the multiple ethnonyms used to label the disease since it made its first apparition in 1495. Díaz de la Isla recounts how initially the blame fell upon the nearest or most problematic enemy, constituting a long chain of accusations that traced the dispersion of the disease back to Naples and to its French assailants, earning it the name of *morbus gallicus*. Years later, Amerinds and *Indianos* (Indianized European colonists) became a more important and problematic other. Reports from the West Indies started being published and the blame for the pox or syphilis fell upon Columbus' crew and upon the invaded Amerinds. In this way, the Columbian theory of the origin of syphilis became established. This theory postulated that syphilis was an Amerindian disease, which Columbus and his crews had brought with them back to Europe.[11]

Beyond its political connotations, the Columbian theory was backed by ample documentary evidence. European physicians attested to the novelty of the venereal disease and recorded its characteristics, treatments, as well as how the disease changed over time.[12] Early missionaries, chroniclers, and historians of the West Indies reported their personal observations on its frequency, ubiquity, and commonality in the Americas.

The first European bearers of the disease, so the legend goes, were Juan de Moguer, as well as Pinzón and the rest of his party, who were among Columbus' crews and participated of the siege of Naples. Oviedo reasoned that the 'passion of the pox' was brought from the Western Indies.[13] He was present at the arrival of Columbus in the first and second trips and was a personal friend of Pinzón until his death in 1514. He witnessed how many members of these crews, those who had 'carnal participation' with Amerind women, fell to the disease and further spread it. Moreover, he worked in Naples some years after the first outbreak, on time to witness the disfiguring sores, the sadness, the crippled, and their raging madness. Finally, he maintained correspondence with Fracastoro, the physician who coined the syndrome with the name syphilis.[14]

On the other hand, Oviedo indicates he saw, through his travels and through the accounts of various personal witnesses, that the pox was a common disease in the Indies. He claimed that the Amerinds did not give it much importance, because the symptoms of the pox were not as severe among them as they were for the European. Las Casas (1951 [1559]), who was highly critical of Oviedo and of most of his opinions, agrees with this attribution of origin, indicating that he was informed that the pox was known from old times to the naturals of the region. Instead they suffered more from the smallpox brought by Europeans (an unrelated disease of viral origin).

Lopéz de Gomara,[15] who participated in the conquest of Mexico, agrees about the origin of the pox in the Indies and its transmission to the Spaniards through sexual intercourse with Amerind women. He also ratifies the commonality and little danger it represented to the Amerinds in contraposition to the smallpox. He indicates that it was the smallpox rather than the sword that caused the greatest fatality among Amerinds and further indicates the Spaniards planted the disease releasing an infected black man, an African slave, who was adopted by the Amerinds, and contaminated them. Later, he adds that smallpox spread was further aggravated by the compulsive Amerindian habit of continuously bathing, and in particular bathing when feverish to wash off diseases.

Amerindian and Medieval Europeans had explanatory frameworks accounting for both diseases and the deaths they produced. From the observations of Lopez de Gomara, it can be concluded that those explanatory frameworks had a counterproductive effect aggravating these diseases: the mortality of smallpox may have been fostered by hygienic measures, which, albeit necessary for controlling bacterial diseases like syphilis, were ineffective against viral diseases like smallpox and measles.

Oviedo attributed the greater resistance of Amerinds to the pox, to the longer exposure to the disease. In his opinion, this early exposure had enabled the development of a thorough knowledge of a wide variety of cures, which could be practiced with ease, given the availability of resources. Moreover, the weather contributed to the effectiveness of the cures.[16]

Oviedo describes a variety of treatments, like eating the flesh of humming birds, but he was particularly emphatic on the curative power of the hard and aromatic wood of the guaiacum (*Guayacum officinale*) and the Palo Santo (*Guayacum sanctum*). Amerindians, he says, drinking and bathing with decoctions from this tree, practicing fasting, sexual abstinence, and strenuous labour for some (30) days recovered easily from this disease without further ado. He claimed to have witnessed the power of the cure at work on many Europeans who recovered from being utterly crippled and devastated with open wounds. As mayor of Santo Domingo, Oviedo had part in the highly profitable business in guaiacum wood, which was being exported to Europe as early as 1510.[17] Fracastoro describes this remedy and acknowledges it as the most effective treatment if accompanied by fasting, strenuous labour, and sexual abstinence.[18]

Fracastoro is widely acknowledged as the Renaissance man who proposed the *Theory of Contagion* and gave syphilis its name, providing one of the earliest medical descriptions of the disease, its origins, and its cures.[19] Like other physicians of his time, Fracastoro recognized syphilis as a contagious venereal disease. He observed that the disease generally attacked genitals first, and he agreed with the general opinion that this was generally the initial site of contagion; for this reason early on the disease had been called *lues venerea* (venereal plague) and *pudengra*.[20]

Advocating for a *Theory of Contagion*, Fracastoro was not breaking away from the medical tradition of his time. Although through the late Middle Ages and the Renaissance the humoral theory of disease predominated, it included a small list of contagious diseases, such as rabies, ophthalmia, scabies, and smallpox, while others, like the black plague, remained contentious. From the early fifteenth century, public health officers considered the black plague as contagious and based on this assumption implemented quarantines and other measures to contain the disease. However, the issue was confounded by the role of flies and rodents as necessary vectors and intermediate hosts of the pathogens. Moreover, humoral medicine placed its interest in the physiological processes of the organism, understood through the dynamics of the four humours: blood, phlegm, vile, and black vile, diseases did not exist as entities in themselves.[21]

Within the humoral theory, contagion was understood as form of poisoning, or as a phenomenon akin to magnetism or to the shock of an electric fish. That is, contagion was regarded as the effect of inert matter or forces on the equilibrium of the body. A remnant from this tradition is the notion of virus, which was the Latin for venom or poison, and in general is still considered as inert because it depends on host cells for its reproduction.[22] More frequently following the miasmatic theory, it was assumed that plagues were caused by miasmas, by air that was out of balance, was too humid or to dry, by air that had acquired properties that did not correspond to its characteristics. In either case, rather than an intruding entity attacking the body, what was transmitted through contagion was a misbalance of the humours, a dysfunction of the body.[23] These notions were tightly associated and confounded with religious notions of moral contamination.[24]

Fracastoro (1863 [1535], 23) recognized that syphilis was unknown to European medicine and credited the reports indicating that the disease was common through the New World. However, he believed that it had spread too quickly, appearing almost simultaneously around the world. In his view, Columbus' crew could not be held responsible for such a swift displacement. Following accepted knowledge of that time, he resorted to astrological reasons to explain the fundamental causes of the plague. According to Fracastoro (1863 [1535], 19), every couple centuries a conjunction of Mars and Saturn brought war and corrupted the air, bringing about propitious conditions that favoured the *growth*, *reproduction*, and *transmission of these semina*, giving rise to plagues, such as syphilis. This description contains the real innovation brought by Fracastoro, although he placed the ultimate causation in an astrological event; his description already acknowledges the biotic characteristics of *semina* and recognizes them as living and proliferating agents, which are transmitted through the air, objects, or corporeal contact and which actively attack the body, causing disease and resisting treatment. It is worth noting the phonetic and semantic proximity between Fracastoro's notion of *semina* and the Taino notion of cemi or zeme, or the Callinago *çemeijn*.

Besides describing the disease as contagious and explaining the astrological reasons for the plague, Fracastoro also presents two mythological accounts for its origins while discussing the source of the knowledge about its cures. Describing the cures, Fracastoro first explains general conditions for treatment, emphasizing in the need to follow a detailed diet, for strenuous activity, and to avoid sexual intercourse. Then Fracastoro describes the mild and strong cures.

Fracastoro places particular emphasis on the use of mercury as a cure for the disease.[25] According to him, the knowledge of these treatments has its origin in the humid forests of the valleys of Syria. A hunter called Ilceo falls to the disease and asks for advice from the nymph Callirhoe. The nymph informs him that he had brought upon himself the wrath of Diana (the moon goddess) for hunting a sacred deer, which he had commemorated to the nymph. Apollo, brother of Diana, in solidarity to his sister, punished Ilceo with the disease, for which nowhere where the sun (Apollo) shines will he find a cure. To be cured Ilceo must entrust himself to the nymphs of metallurgy, who dwell in deep caverns and know all the secrets of the alloys. Ilceo follows the advice and receives treatment from the Nymph Lipare, who is in charge of gold, silver, and copper alloys; she baths him in quicksilver and cures him.

In his final chapter, Fracastoro turns his attention to 'forests of the New World' and makes a brief allusion celebrating D'Anghiera's poetic work, the *Decades of the New World*, to introduce his more modest piece dedicated to the guaiacum tree and the cure obtained from it. First, he describes the guaiacum tree and the mode to prepare and administer the cure, as well as the rigorous diet to which the inhabitants of Hispaniola submit themselves to insure its success. Then he writes about the way in which the cure came to be known:

> Fracastoro starts with an elegy to the navigations of the Spanish explorers. However, upon arriving to an island in the New World, they shoot upon a flock of blue birds. One of the birds tells them they have killed birds consecrated to the Sun, a crime for which they will be punished with discord and an unknown disease. The crew meets the local inhabitants, and after the formalities come to discover the disease. The chief of the island tells the explorers that the inhabitants are the descendants of the race of Atlantis, who were punished by the vengeful sun, *Apollo,* for their impiety. A shepherd called Syphilos had compared the few animals of the heavens with the many owned by the king and proposed to adore the king instead of the gods. To calm the gods Syphilos was to be sacrificed, but Apollo forgives him, replacing him with a white calf. From the site of the sacrifice sprouts the guaiacum tree, the cure to the disease.[26]

By the end of the century, the effectiveness of the cure decreased and it was widely discredited as a superstitious practice. European physicians inclined to the cures with mercury. Defenders of the qualities of the guaiacum argued that the problem lied in the lack of freshness of the species; however even in the Indies a similar situation was being observed. The discredit of guaiacum may have been unjustified, as methyl extracts of *Guayacum* spp. do have strong anti-bacterial properties effective against gram-negative bacteria.[27] It is possible that through prolonged and unregulated use, *T. pallidum* developed a resistance to the active components in guaiacum. In much the same way, it has developed a resistance to strong-line antibiotics, like azitromicine.[28]

The failure of guaiacum questioned the value of Amerindian medicine in general and seemingly refuted the 'Columbian theory'.[29] Nevertheless, both of the main treatments and the associated myths recorded by Fracastoro were interpretations of Amerindian treatments for syphilis, codified in the myth of the Sun and the Moon. In the case of guaiacum even though Fracastoro adapts the myth to match European expectations, he acknowledges its Amerindian origin. Likewise, the *Theory* of *Contagion* developed by Fracastoro reproduces multiple aspects of coetaneous Amerindian understandings of the causes of disease. Furthermore, the very notion of *semina* is phonetically with Amerindian terms for potential pathogens such as zeme and *çemeijn*. Also suggestive of the conceptual borrowing was the poor reception of Fracastoro's *Theory of Contagion*, the fact that he himself regarded it as merely an addition to the *Humoral Theory*. The resistance to this notion is understandable considering Christian religious beliefs, which ever since Ockham in the fourteenth century had dispelled such forms of agency.[30]

11.2 Seventeenth- to Twentieth-Century Syphiloids, the *Ontological Theory of Disease*

However, the debate on the origin of syphilis continued. It became more complex and multilayered. Towards the middle of the sixteenth century, Thevet described a Brazilian disease, with symptoms that coincided considerably with those of the pox, which the Tupinamba Amerindians called *pians*. Based on his opinions on Tupinamba sexuality, Thevet concluded that the disease was venereal and that it was the original disease from which the pox had sprouted.[31] Later on, however, he described a Persian leper-like disease called *lubard* or *bulich*, which, presenting remarkable similarities with *pians*, was clearly non-venereal, leading him to question his previous assumptions regarding the Brazilian disease.[32] A century later, Du Tertre (1954 [1654], 481) records *pians* as a name given to the pox in the Antilles. He noted, however, that some cases of *pians* were clearly not of venereal origin, and further suggested that living conditions played a role in the transmission. About the same time Breton has *yáya* as the Callinago name for both *vérolle* (Fr. for pox) and *pians* (Fr. for yaws).[33] Twenty years later, yaws had become a common illness amongst British seaman. Sydenham (1848), after analysing the characteristic symptoms associated with the diseases, concluded yaws was caused by the same disease entity as the pox. On the basis of verbal testimonies describing how the disease broke out on-board slave ships before their arrival to America, he also concluded that it was of African provenance.[34] The analysis of syphilis and yaws was one of the early foundations of Sydenham's nosology (the characterization of diseases according to a botanic-like classification of their symptoms). It also played a pivotal role in his ontological theory of disease, which argued that diseases were caused by entities that had an existence independent of the body and its humours.

Parallel to these observations, from the middle of the seventeenth to the nineteenth century, a number of syphiloids, pathologies closely related to syphilis, yet differing in their epidemiology, were reported to have terrorized rural Europe. The Scottish knew the disease as *sibbens*, the Irish as button scurvy, the Norwegian as *radesgye*, and the Swedish as *saltfluss*. In the German–Danish conflict regions of Schleswig-Holstein and Jutland, the disease was known *dithmarsh evil*, in France as the disease of Chavane-Lure and as the *pian de Nérac*, as for the Italians it was the *facaldine*, for the Greeks *spirocolon*, the Czechs called it the disease of *Brün*, and the Bosnians *skerljebo*.[35] Syphiloids, moreover, were being reported from European colonies overseas, ranging from Canada to Africa to Papua New Guinea.[36]

Syphiloids were regarded as 'social' rather than venereal diseases. They often affected children as highly contagious rashes. Their transmission was attributed in some instances to sharing eating utensils and smoking pipes, and in other cases to skin-to-skin contact. Eroding the palate and the nose and transforming the extremities, syphiloids caused terrible deformations. Nonetheless, their non-venereal character spared the diseased from the stigma of syphilis. The rural endemic character and the various names for syphiloids implied the possibility that these diseases were already present in Europe before the Encounter with America. Nonetheless, syphiloids were endemic of poor rural areas, associated with lack of education and precarious hygienic conditions, which compared negatively to the prosperous and creative, though promiscuous, environments associated with syphilis. While syphilis stigmatized the individual, its endemic non-venereal relatives stigmatized societies. In the context of the consolidation of the European nation-states and of Romantic Nationalism, the debate on the origin of syphilis acquired further biopolitical connotations conflating not only concerns about sexuality but also about urbanization, hygiene, cosmopolitanism, and ways and qualities of knowing. In time, the origin of syphiloids was attributed to extra-European peoples. Thus, while syphiloids from Eastern Europe were blamed on the invading Turkish Empire, the introduction of sibbens into Scotland was blamed on mariners, drawing a connection to Caribbean *yaws*.[37]

By the middle of the nineteenth century, syphiloids had been eradicated from most of Europe, presumably as an effect of the depopulation and the hygienization campaigns brought by modernity. Syphilis, however, could not be eradicated, and various authors inclined for the pre-Columbian theory of the origin of syphilis. They attributed the apparent discovery of a new disease at the end of the sixteenth century to a previous misclassification and confusion with leprosy, citing earlier instances of a venereal or congenital leprosy and treatments containing mercury—characteristics which do not correspond with the modern understanding of Hansen's disease.[38] Indeed, it is possible that *leprosy* was formerly a broader term as early descriptions of syphilis classified it as a kind of leprosy.[39] However, as has been pointed by Crosby, rather than attempting to differentiate syphilis from leprosy, physicians like Diaz de la Isla draw on their resemblances, as is often the case when dealing with novel phenomena.[40] Moreover, the cited descriptions from venereal or congenital leprosy are generic and ambiguous, corresponding equally well to various other venereal diseases, some of which also respond to treatments with mercury.[41]

11.3 Twentieth Century, the Entangled Debates of *Treponema pallidum*

At the turn of the twentieth century, the spirochaete *Treponema pallidum* was identified as the causal agent of syphilis. Overtime, however, it became evident that *Treponema* also caused yaws (*T. p. pertenue*), bejel (*T.p. endemicum*, a non-venereal 'Arabian endemic syphilis' described in the 1920s), and pinta (*T.p.* carateum), a mild skin disease endemic of Central and South America, which until then had seemed unrelated to syphilis.[42] These diseases became identified as the non-venereal treponematoses of developing countries. Nonetheless, the *treponemata* causing the four diseases cannot be differentiated morphologically, serologically, or immunologically from one another. This led Hudson to propose the Unitarian theory, arguing that the different treponematoses are in fact a single treponematosis: a single disease differently expressed in response to different conditions of the environment (humidity, temperature) and to the opportunities of transmission available. Thus, in yaws, *Treponema* took advantage of the insalubrious conditions of the humid tropical weather, while bejel took advantage of water scarcity in the desert. Hudson argued further that the appearance of venereal syphilis must have been closely associated with the process of urbanization experienced in the West during the Renaissance.[43] Thus, *Treponema* and its victims appeared as the tools and subjects of environmental determinism.

Nonetheless, treponemata can be differentiated by animal sensitivity, the particular ways in which treponemal infections are propagated and expressed in different animals.[44] *T. p. pertenue*, *endimicum*, and *pallidum* can be propagated in mice, hamsters, guinea pigs, rabbits, and primates. However, *T.p. pallidum* only causes infections that resemble syphilis in rabbits and primates. Finally, *T. pallidum carateum* can only be propagated in primates. The features of each pathogen are retained even after 500 passages in animals, indicating that they are not responses of the same species to different environments.[45]

Meanwhile, Hackett (1963) attempted to substantiate the pre-Columbian theory on epidemiological models, inversely correlating treponemal infectiousness with population density. Based on modern distributions of treponemal disease, he concludes that all treponematoses evolved from pinta, the ancient *Treponema* isolated in low-density populations of the Americas. Elsewhere in the world, the disease would evolve into yaws, later in desert regions into bejel, finally turning into syphilis. With its venereal character, he argues, syphilis would have been favoured only by urbanization, and thus must have evolved with Western civilization since its Sumerian birth. Such conclusions presuppose a lineal and unidirectional evolution of both society and of the treponemal pathogens, either of which are difficult to justify on biological or historical grounds. In particular, given that due to the European colonialist expansion from the fifteenth-century onwards, present distributions of treponemal diseases cannot be extrapolated into the past.

Arguably, claims to the Columbian and pre-Columbian theories have been substantiated by paleopathological evidence.[46] Nonetheless, a recent review challenged all the reports of non-American pre-Columbian paleopathological evidence, indicating that either the certainty of diagnosis of treponemal disease and/or the dating was dubious.[47] Those cases in which there could be a confident diagnosis of a treponemal disease, and which were carbon dated, once calibrated, presented temporal ranges overlapping with the early sixteenth century. Meanwhile, the early presence of treponematoses in the Americas is well attested at multiple sites.[48] The earliest evidence of treponematoses is found in Colombia in a high plateau context (2600 m above sea level), in which a male and a female from circa 3075 B.C.[49] present signs of advanced treponemal infection (tertiary syphilis) and are serologically positive to *Treponema pallidum*, and a female from circa 2080 B.C.[50] presents clear signs of congenital malformations specifically associated with syphilis, which are generally accepted as differentiating grounds that single out the presence of this particular treponematosis in an archaeological context.[51]

Historical records on the other hand evidence that the most striking feature of the pox epidemic at the end of the fifteenth century was its venereal character. With that restricted mode of contagion, and the assistance of growing imperia, the disease spread through both urban and more rural areas of Europe and Asia. Endemic non-venereal syphiloids or treponematoses were only identified outside America 70 years after the Encounter and in Europe a century and a half later. Notably Amerindian ethnohistorical sources are the earliest reports distinguishing varieties of the disease, suggesting that the disease had already differentiated in America before the arrival of Columbus.

11.4 Current Understandings of *Treponema pallidum*

Many species of *Treponema* (lit. *Threadlike*) are symbiotic guests, inhabiting the bodies of other organisms. Many of them are mutualists that collaborate with the digestion of vegetable matter and woody fibres, in the hindgut of cockroaches and termites, the rumen of cows and sheep, the intestines of swine, and even in our mouths.[52] Through this long-standing and mutually beneficial symbiosis, *Treponema* have increasingly simplified their genome, becoming ever more dependent on their hosts. Following this trend, some *Treponema* have become pathogenic, the sources of disease like human periodontitis (*T. denticola*), a venereal disease in rabbits *(T. paraluis-cuniculi)*,[53] and a group of complex syndromes affecting humans caused by *T. pallidum*.

Turning into a parasitic pathogen has implied a further simplification of the genome of *T. pallidum*, which in comparison with free-ranging spirochetes has been reduced sixfold to only about 900 genes, delegating multiple cell cycles and the processing of numerous building materials into its hosts.[54] In this way they have become obligate parasites, with very particular requirements for optimal growth: moderately low temperature (33–35 °C), glucose or maltose as unique sources of energy, and despite finding atmospheric oxygen toxic, tiny concentrations of the latter for their metabolic functions.[55] These constrictions make the culturing of

Treponema pallidum in vitro extremely difficult. Despite these constraints, *T. pallidum* is highly labile and has diversified into at least four subspecies.

Three subspecies of *Treponema pallidum—pertenue, endemicum, and pallidum*—are very unwelcome and invasive guests of humans as they produce three terrible diseases, yaws, bejel, and syphilis (respectively), while *Treponema (pallidum) carateum* produces pinta, a milder skin disease. Recent genomic analysis confirms their close proximity and allows the discrimination of all but one variety, *T. [pallidum] carateum*, for which samples can no longer be found.[56] The degree of genetic variation indicates that the subspecies differentiated more than 500 years ago, that is, before the contact.[57]

11.4.1 Yaws

T.p. Pertenue, the causal agent of yaws, spreads particularly among children by skin-to-skin contact, in particular under high humidity and low sanitary conditions.[58] Coastal peoples are particularly exposed, even if otherwise enjoying healthy conditions, which do not indicate poverty.[59] Today, *T.p. pertenue* is endemic to rural coastal zones lying on the Equatorial belt: tropical Africa, Indonesia, South America, and the Caribbean. However, it has the same low temperature requirements as other *T. pallidum*—their optimum is slightly below body temperature.[60] Thus, the conditions of low temperature and high humidity encountered in many temperate regions are actually optimal for its development. Furthermore, yaws is as rare in modern urban tropical contexts as it is in modern temperate regions. The only marked difference with the neighbouring rural tropical areas is the use of shoes and dry clothing, restraining its transmission.[61]

The first expressions of yaws are highly infectious dermal granulomatous (grain-like) lesions in the colder regions of the body, which tend to be the lower extremities, particularly the forelegs. These lesions, steaming with infective parasites, can become ulcerated, facilitating their spread. In a secondary stage, the wounds may heal and recur, spreading through the lymphatic system and affecting all the surface of the body, even the palms of hands and feet. At this stage, the disease also attacks the joints, producing swelling and pains linked with the growth of the bacteria in the attachment sites of the muscles on the bones and the resulting calcification of cartilages. Finally, in the tertiary stage, it is assumed that clusters of bacteria turn highly invasive, attacking skin and bone. This is doubtful, given that at this point the degree of infectiousness is minimal. There are, however, a series of violent reactions of the immune system, which generate crater-like (gummatous) lesions as well as superficial and profound transformations of the bone which becomes hypervascularized to augment the blood flow to the region, cleanse, and augment the temperature at the extremities.[62] This produces the characteristic sabre shin tibia and other bone transformations.

It is widely assumed that the noxious effects of yaws are limited to the skin and skeletal systems. There is, however, strong evidence indicating that at least American varieties of yaws can produce damages in internal organs and affect the central nervous system and the sensorium. It can produce peripheral neuropathies, lesions in the crystalline of the eyes, tinnitus, and recursive deafness.[63] Moreover, there is evidence indicating that these varieties of yaws can produce congenital complications.[64]

11.4.2 Bejel

T.p. endemicum is transmitted solely during youth, through mouth-to-mouth contact or through the sharing of contaminated eating and drinking utensils. Like yaws it has three stages; however, in the first stage it is confined to the mucous regions, in particular the mouth and palate. During the secondary stage, it becomes extended to moist regions of the skin, like the axillae and sometimes the extremities. Finally, during the tertiary stage, it also compromises the bones, producing deep deformations, frequently eating the palatal zone away, affecting the extremities but usually sparing the hands and feet, and rarely affects the nervous system or the sensorium.[65] Given the particularities of bejel, it is well adapted to dry environments and low sanitary conditions. Nomadic peoples in dry regions are at a high risk because the limitations on water supplies force the sharing of drinking vessels.[66]

11.4.3 Syphilis

The fearsome *Treponema pallidum* subsp. *pallidum* is the causal agent of syphilis, otherwise known as the 'pox', the 'kiss of death', or the 'great imitator'. In early stages, it seriously compromises the skin, and in later stages, the musculoskeletal system. However, it is generally thought that syphilis differs dramatically from other treponematoses in three respects: the disease is almost exclusively venereal, it is congenital, and it is a multisystem disease which compromises almost every organ of the body, namely, the heart, the sensorium, and the nervous system. The latter two are implicated from the very early stages, perhaps as little as 18 hours after infection.[67]

Syphilitic infection is evidenced after three weeks of contagion by a painless callous chancre. The chancre disappears quickly, lending host and guest between six months and seven years of unnoticed infectious ease. Through this time, *T. p. pallidum* travels swiftly through the blood and the lymph, establishing colonies in tissues all over the body. After the interim, systemic infection starts—this is the second act. The pain begins with headaches, nausea, a generalized fever, and a highly symmetrical rash that may be subtle or extend to every surface of the skin and then disappear. Then, as if it was following every possible route of somatization, it hits those faculties which are most prized to the host: Beethoven's ear; the eyes of Van Gogh, Baudelaire, and Maupassant; Nietzsche's nerves, thoughts, and speech; a beautiful maid's skin, bones, articulations, the heart, respiration, and the stomach.[68] Its expressions at this stage are varied; it closely resembles a wide variety of afflictions that can easily mislead physicians. As said William Osler, father of modern medicine:

> Know syphilis in all its manifestation and relations, and all other things clinical will be added on to you.[69]

During secondary syphilis, the infection becomes latent, pains, and sorrows—signs of completely unrelated afflictions—appear and disappear intermittently like different actors; the host appears to be cured and to relapse, reiteratively. Deceitful, syphilis confers the host a reputation for hypochondria.

Finally, after an indeterminate period of time, *T. p. pallidum* might advance into a third stage, compromising the functional integrity of the skin and bones (gumma), internal organs, the heart, and the nervous system. The effects of syphilis on the nervous system have been regarded with a dreaded fascination, combining the recognition of genius and the fear of madness.[70] Following a time of deep disinhibition, mystical insights, and sometimes the composition of 'odes to joy' and equivalent masterworks comes the dreaded paralytic dementia or the paralysis of the insane, the syndrome of PARESIS. This acronym stands for the disturbances, abnormalities, hyperactivity, and impairment it produces in *p*ersonality, *a*ffect, *r*eflex, *e*ye, *s*ensorium, *i*ntellect, and *s*peech. In a process that can take from a few months to up to 30 years, the host raves, rolls, and becomes increasingly paralyzed. Again, this stage is intermittent with episodes of insightful sanity mixed up with periods of ululating violence and petrified stasis.

11.4.4 Pinta or Carate

Endemic to the Caribbean, Central, and South America, *Treponema (pallidum) carateum* produces a milder skin affliction. Its mode of propagation remains unknown, although it is thought to involve sweat, perhaps prolonged and intimate contact, and repeated skin-lesion contact. It is suggested that scanty clothing promotes its transmission.[71] Unlike the other treponematoses it has only two stages. It starts with a small lesion in the extremities, which might produce one or two satellite lesions on nearby lymphatic nodes. Like the first lesions, these steam with infectious spirochetes. As these lesions heal, they leave hyperkeratosis, a thickening of skin layers, or alternatively decolorized or hyperpigmented blue marks of the same kind seen in tertiary syphilis and occasionally in yaws. In the secondary stage, lesions spread through the skin, near articulations or in places where the bones approach the surface. As these lesions heal, they can produce widespread and highly symmetrical mottled patterns, which feature frequently in jokes that remark on the resemblance with the jaguar. The infection remains latent, reappearing periodically, but does not progress into a third phase. Late infection with *T.p. carateum* does not compromise the bones; and hence it has been classified as milder and somewhat more distantly related to the other human *T. pallidum*. However, like *T.p. pallidum* and some strains of *T.p. pertenue*, *T.p. carateum* can be detected in cerebrospinal fluid. And some psychiatric examinations have suggested mild confusion and peripheral neuropathies such as numbness or hyperactivity of the senses, motor, and/or autonomic systems.[72]

From the 1950s to the 1970s, it was observed that after the secondary stage of pinta, the blood and cerebrospinal fluid developed an antigenic reaction that immobilized or restrained spirochetes of *T.p. carateum*, as well as those of *T.p. pallidum*, which it further eliminated. This reaction was understood as the reason for the blockade to the third phase of the disease and its damaging consequences, as well as the source for cross-immunity, evidenced in resistance to other varieties of *Treponema pallidum*.[73] Syphilitics present a similar but less effective antigenic reaction that only slows down spirochetes. Cross-infection experiments have shown that after the tenth week, persons with yaws exhibit resistance to pinta, and similarly persons with pinta after the outset of second stage are highly resistant to yaws. On the other hand, syphilitics are readily infected by *T. carateum*. Meanwhile it is virtually impossible to infect with syphilis those who have pinta.[74] The methods of this line of research are questionable on ethical grounds; nevertheless, the findings indicate a promising route towards the development, or rather recognition, in pinta of something akin to a vaccine against the fearsome syphilis.[75] These features explain the cultural value attached to these conditions in various Amerindian contexts. However, despite their recognition of its inoffensiveness and important prospective as a vaccine, these researchers classified pinta as just another endemic disease, provoking a 'repulsive aspect' associated with conditions of 'lack of education, poverty, crowded living spaces and lack of hygiene'.[76] The WHO promoted eradication campaigns, which may have led to its extinction.

Notes

1. Hayden 2003.
2. Tramont 1995; Waugh 1982; Abraham 1944; Hudson and Morton 1996.

3. Hudson and Morton 1996.
4. Crosby 2003, 151.
5. Reed 2004.
6. Hayden 2003.
7. Morton 1990a.
8. Ehrlich 1913; Porter 2006, 87.
9. von Krafft-Ebing 1898, Vol. 2, 20; Clark 2013; Freyre 1986.
10. Hayden 2003.
11. Díaz de Isla 1539, 2.
12. Díaz de Isla 1539; Fracastoro 1863 [1535], 1930 [1546], 135–141.
13. Fernández de Oviedo 1959 [1535]T 1, lb. 2, cp 7, fl 9; cp 13, fl18; cp 14.
14. Fracastoro 1930 [1546], xxi.
15. Excerpt in Girtanner 1789, 881–886.
16. Fernández de Oviedo 1959 [1535], Lb 2, C10, fl 92.
17. Hudson and Morton 1996.
18. Fracastoro 1863 [1535], 94–99.
19. Tramont 1995; Waugh 1982; Abraham 1944; Hudson and Morton 1996; Morton 1990b.
20. Fracastoro 1930 [1546], 153.
21. Carmichael 1991.
22. Lewis and Short 1879.
23. Nutton 1983.
24. Harrison 2013.
25. Fracastoro 1863 [1535], 74–76.
26. Fracastoro 1863 [1535], 94–99.
27. Castillo-Juáreza et al. 2009.
28. LaFond and Lukehart 2006; Stamm 2010.
29. Reyes Equiguas 2009.
30. Tachau 1988.
31. Laborie and Lestringant 2006, 211–214.
32. Thevet 1575, 374.
33. Breton 1666, 291, 399.
34. Due to incubation times, it is more probable that such testimonies indicate that slaves were infected while in captivity. Moreover, the purported association between sibbens and yaws further suggests that European slave ship crews were vectors of the disease, regardless of origin and destination.
35. Morton 1967; Morton and Rashid 2001; Hudson 1961.

36. Rayer 1835.
37. Morton 1967.
38. Brothwell 1961.
39. Díaz de Isla 1539, iv.
40. Crosby 1969; Díaz de Isla 1539, v.
41. Baker et al. 1988.
42. Hudson 1958.
43. Hudson 1958.
44. Turner and Hollander 1957.
45. Turner and Hollander 1957.
46. A review in de Melo et al. 2010.
47. Harper et al. 2011.
48. Powell and Cook 2005.
49. Carbon dated to 5025±40 B.P. (Gr. No. 14.447, Col 7.952).
50. Carbon dated to 4030±80 B.P. (Gr. No. 12930 Col. 477).
51. Correal 1990, 209–211; Burgos et al. 1994.
52. Husseneder 2010; Berlanga et al. 2010; Stanton and Canale-Parola 1980; Norris et al. 2006.
53. Smith and Pesetsky 1967.
54. Margulis 2004.
55. Norris et al. 2001.
56. Norris et al. 2006; Centurion-Lara et al. 2006; Harper et al. 2008.
57. Šmajs et al. 2012.
58. Antal et al. 2002.
59. Drew 2009.
60. Norris et al. 2001.
61. Antal et al. 2002.
62. Antal et al. 2002.
63. Román and Román 1986; Smith et al. 1971; Black et al. 2007.
64. Román and Román 1986.
65. Norris et al. 2006.
66. Antal et al. 2002.
67. LaFond and Lukehart 2006.
68. Hayden 2003.
69. Silverman et al. 2008, 437.
70. Hayden 2003.
71. Varela 1957; Antal et al. 2002.
72. Smith et al. 1971.
73. Varela 1957, 1963.

74. Medina 1979.
75. Thatcher 1969.
76. Medina 1979; Varela 1957, 1963.

Bibliography

Abraham, J.J. 1944. The early history of syphilis. *British Journal of Surgery* 32 (126): 225–237.

Antal, Georg Michael, Sheila A. Lukehart, and André Z. Meheus. 2002. The endemic treponematoses. *Microbes and Infection* 4 (1): 83–94.

Baker, Brenda J., George J. Armelagos, Marshall Joseph Becker, Brothwell Don, Andrea Drusini, Marie Clabeaux Geise, Marc A. Kelley, Iwataro Moritoto, Alan G. Morris, George T. Nurse, Mary Lucas Powell, Bruce M. Rothschild, and Shelley R. Saunders. 1988. The origin and antiquity of syphilis: Paleopathological diagnosis and interpretation [and comments and reply]. *Current Anthropology* 29 (5): 703–737. https://doi.org/10.2307/2743609.

Berlanga, M., B.J. Paster, and R. Guerrero. 2010. The taxophysiological paradox: Changes in the intestinal microbiota of the xylophagous cockroach cryptocercus punctulatus depending on the physiological state of the host. *International Microbiology* 12 (4): 227.

Black, R.J., W.P.R. Gibson, and J.W.R. Capper. 2007. Fluctuating hearing loss in West African and West Indian racial groups: Yaws, syphilis or Menière's disease? *The Journal of Laryngology and Otology* 96 (09): 847–855.

Breton, Raymond. 1666. *Dictionaire français-caraïbe*. Auxerre: Gilles Bovqvet, Imprimeur ordinaire du Roy. Original edition, 1666.

Brothwell, Don. 1961. The palaeopathology of early British man: An essay on the problems of diagnosis and analysis. *The Journal of the Royal Anthropological Institute of Great Britain and Ireland* 91 (2): 318–344. https://doi.org/10.2307/2844418.

Burgos, J.D., G. Correal-Urrego, and C. Arregoces. 1994. Treponematosis en restos oseos preceramicos de Colombia. *Revista de la Academia Colombiana de Ciencias Exactas, Físicas y Naturales* 19: 273–281.

Carmichael, Ann G. 1991. Contagion theory and contagion practice in fifteenth-century milan. *Renaissance Quarterly* 44 (2): 213–256. https://doi.org/10.2307/2862709.

Castillo-Juáreza, Israel, Violeta Gonzáleza, Héctor Jaime-Aguilara, Gisela Martíneza, Edelmira Linaresb, Robert Byeb, and Irma Romeroa. 2009. Anti-Helicobacter pylori activity of plants used in Mexican traditional medicine for gastrointestinal disorders. *Journal of Ethnopharmacology* 122 (2): 402–405. https://doi.org/10.1016/j.jep.2008.12.021.

Centurion-Lara, A., B.J. Molini, C. Godornes, E. Sun, K. Hevner, W.C. Van Voorhis, and S.A. Lukehart. 2006. Molecular differentiation of *Treponema pallidum subspecies*. *Journal of Clinical Microbiology* 44 (9): 3377.

Clark, Hannah-Louise. 2013. Civilization and syphilization: A doctor and his disease in colonial Morocco. *Bulletin of the History of Medicine* 87 (1): 86–114.

Correal, U. 1990. Aguazuque: Evidencias de cazadores, recolectores y plantadores en la altiplanicie de la Cordillera Oriental.

Crosby, Alfred W., Jr. 1969. The early history of syphilis: A reappraisal. *American Anthropologist* 71 (2): 218–227. https://doi.org/10.2307/669854.

Crosby, Alfred W. 2003. *The Columbian exchange: Biological and cultural consequences of 1492*, vol. 2. Westport: Greenwood Publishing Group.

de Las Casas, Bartolome. 1951 [1559]. *Historia de las Indias*. Ed. Agustín Millares Carlo. México D.F.: Fondo de Cultura Económica. Original edition, 1559.

de Melo, F.L., J.C.M. de Mello, A.M. Fraga, K. Nunes, and S. Eggers. 2010. Syphilis at the crossroad of phylogenetics and paleopathology. *PLoS Neglected Tropical Diseases* 4 (1):e575 (1–11). https://doi.org/10.1371/journal.pntd.0000575.

Díaz de Isla, Ruy. 1539. *Tractado co [n] tra el mal serpentino que vulgarmente en España es llamado bubas*. Lisboa: (H)ospital de Todos los Santos.

Drew, Rose. 2009. From Saladoid to Taino: Human behavior from human remains in the Greater Antilles. *Bulletin of the Peabody Museum of Natural History* 50 (1): 167–185.

Du Tertre, Jean-Baptiste. 1954 [1654]. *Histoire générale des isles de S. Christophe, de la Guadeloupe, de la Martinique et autres dans l'Amérique*: chez Jacques Langlois: et Emmanuel Langlois.

Ehrlich, Paul. 1913. Address in pathology, on chemiotherapy. *The British Medical Journal* 2 (2746): 353–359.

Fernández de Oviedo, Gonzalo. 1959 [1535]. Historia general y natural de Indias. In *Biblioteca de Autores Españoles*. Madrid: Biblioteca de Autores Españoles. Original edition, 1935.

Fracastoro, Hieronimus. 1863 [1535]. *La Syphilis: poema latino/de Gerónimo Fracastor; traducido al castellano e ilustrado con notas por Luis María Ramírez y de las Casas-Leza*. Madrid: Imprenta de de José M. Ducazal.

Fracastoro, Girolamo. 1930 [1546]. *Hieronymi Fracastorii De contagione et contagiosis morbis et eorum curatione: libri III*. Trans. Haven Emerson, *History of medicine series*. New York: Putnam.

Freyre, Gilberto. 1986. *Casa-grande e senzala*, vol. 351: Berkeley: University of California Press.
Girtanner, C. 1789. *Abhandlung über die venerische Krankheit*. Gottingen: Johann Christian Dieterich. http://books.google.com/.
Hackett, Cecil John. 1963. On the origin of the human treponematoses: Pinta, yaws, endemic syphilis and venereal syphilis. *Bulletin of the World Health Organization* 29 (1): 7.
Harper, K.N., P.S. Ocampo, B.M. Steiner, R.W. George, M.S. Silverman, S. Bolotin, A. Pillay, N.J. Saunders, and G.J. Armelagos. 2008. On the origin of the treponematoses: A phylogenetic approach. *PLoS Neglected Tropical Diseases* 2 (1): e148.
Harper, Kristin N., Molly K. Zuckerman, Megan L. Harper, John D. Kingston, and George J. Armelagos. 2011. The origin and antiquity of syphilis revisited: An appraisal of Old World pre-Columbian evidence for treponemal infection. *American Journal of Physical Anthropology* 146 (S53): 99–133. https://doi.org/10.1002/ajpa.21613.
Harrison, M. 2013. *Contagion*. New Haven: Yale University Press.
Hayden, D. 2003. *Pox: Genius, madness, and the mysteries of syphilis*. New York: Basic Books.
Hudson, Ellis Herndon. 1958. The treponematoses – or treponematosis? *British Journal of Venereal Diseases* 34 (1): 22.
———. 1961. Endemic syphilis—heir of the syphiloids. *Archives of Internal Medicine* 108 (1): 1.
Hudson, Margaret M., and Robert S. Morton. 1996. Fracastoro and syphilis: 500 years on. *The Lancet* 348 (9040): 1495–1496.
Husseneder, C. 2010. Symbiosis in subterranean termites: A review of insights from molecular studies. *Environmental Entomology* 39 (2): 378–388.
Laborie, Jean-Claude, and Frank Lestringant. 2006. *Histoire d'André Thevet Angoumoisin, cosmographe du roy, de deux voyages par luy faits aux Indes australes, et occidentales*. Paris: Librairie Droz.
LaFond, R.E., and S.A. Lukehart. 2006. Biological basis for syphilis. *Clinical Microbiology Reviews* 19 (1): 29.
Lewis, Charlton T., and Charles Short. 1879. *A latin dictionary*. Oxford: Clarendon Press. Perseus Digital Library Project. Ed. Gregory R. Crane. Tufts University.
Margulis, Lynn. 2004. On syphilis & Nietzsche's madness: Spirochetes awake! *Daedalus* 133 (4): 118–125.
Medina, Rafael. 1979. Pinta: Una treponematosis endemica de las Americas. *Boletin de la Oficina Sanitaria Panamericana* 86 (3): 69–79.

Morton, R.S. 1967. The sibbens of Scotland. *Medical History* 11 (04): 374–380.

———. 1990a. Syphilis in art: An entertainment in four parts. *Genitourinary Medicine* 66 (1–4): 33–40, 112–123, 208–221 & 280–294.

———. 1990b. Syphilis in art: An entertainment in four parts. Part 1. *Genitourinary Medicine* 66 (1): 33–40.

Morton, R.S., and S. Rashid. 2001. "The syphilis enigma": The riddle resolved? *Sexually Transmitted Infections* 77 (5): 322.

Norris, S.J., D.L. Cox, and G.M. Weinstock. 2001. Biology of treponema pallidum: Correlation of functional activities with genome sequence data. *Journal of Molecular Microbiology and Biotechnology* 3 (1): 37–62.

Norris, Steven, Bruce Paster, Annette Moter, and Ulf Göbel. 2006. The genus Treponema. In *The Prokaryotes*, ed. Martin Dworkin, Stanley Falkow, Eugene Rosenberg, Karl-Heinz Schleifer, and Erko Stackebrandt, 211–234. New York: Springer.

Nutton, Vivian. 1983. The seeds of disease: An explanation of contagion and infection from the Greeks to the renaissance. *Medical History* 27 (1): 1–34.

Porter, Roy. 2006. What is disease. In *The Cambridge history of medicine*, ed. Roy Porter. Cambridge: Cambridge University Press.

Powell, Mary Lucas, and Della Collins Cook, eds. 2005. *The myth of syphilis: The natural history of treponematosis in North America*. Gainesville: University Press of Florida.

Rayer, Piere. 1835. *Traité théorique et pratique des maladies de la peau*, vol. 3. Paris: Chez J.-B. Baillière.

Reed, C.S. 2004. The codpiece: Social fashion or medical need? *Internal Medicine Journal* 34 (12): 684–686.

Reyes Equiguas, Salvador. 2009. El huauhtli y la cotidianidad nahua. *El mundo indígena desde la perspectiva actual* 3 (18).

Román, Gustavo C., and Lydia N. Román. 1986. Occurrence of congenital, cardiovascular, visceral, neurologic, and neuro-ophthalmologic complications in late yaws: A theme for future research. *Reviews of Infectious Diseases* 8 (5): 760–770.

Silverman, Mark E., T.J. Murray, and Charles S. Bryan, eds. 2008. *The quotable osler*. Philadelphia: ACP Press.

Šmajs, David, Steven J. Norris, and George M. Weinstock. 2012. Genetic diversity in *Treponema pallidum*: Implications for pathogenesis, evolution and molecular diagnostics of syphilis and yaws. *Infection, Genetics and Evolution* 12 (2): 191–202.

Smith, J. Lawton, and Barry R. Pesetsky. 1967. The current status of treponema cuniculi. Review of the literature. *The British Journal of Venereal Diseases* 43 (2): 117–127.

Smith, J., N. Lauton, J. David, S. Indgin, C.W. Israel, B.M. Levine, J. Justice Jr., J.A. McCrary 3rd, R. Medina, P. Paez, and E. Santana. 1971. Neuro-ophthalmological study of late yaws and pinta. II. The caracas project. *British Medical Journal* 47 (4): 226.

Stamm, Lola V. 2010. Global challenge of antibiotic-resistant treponema pallidum. *Antimicrobial Agents and Chemotherapy* 54 (2): 583–589. https://doi.org/10.1128/aac.01095-09.

Stanton, T.B., and E. Canale-Parola. 1980. Treponema bryantii sp. nov., a rumen spirochete that interacts with cellulolytic bacteria. *Archives of Microbiology* 127 (2): 145–156.

Sydenham, Thomas. 1848. *The works of Thomas Sydenham. M.D.* Edited by William Alexander Greenhill, R. G. Latham, Charles Adlard, James Adlard and publisher Sydenham Society, *Publications of the Sydenham Society*. London: Printed for the Sydenham Society.

Tachau, K.H. 1988. *Vision and certitude in the age of Ockham: Optics, epistemology, and the foundations of semantics, 1250–1345*. Vol. 22. Leiden: Brill Academic Pub.

Thatcher, R.W. 1969. The search for a vaccine for syphilis. An epidemiological approach. *British Medical Journal* 45 (1): 10.

Thevet, André. 1575. *La cosmographie vniverselle*. Paris: Guillaume Chaudiere.

Tramont, E.C. 1995. Syphilis in adults: From Christopher Columbus to Sir Alexander Fleming to AIDS. *Clinical Infectious Diseases* 21 (6): 1361–1369.

Turner, T.B., and D.H. Hollander. 1957. Biology of the Treponematoses. *Monograph Series. World Health Organization* (35): 3.

Varela, Gerardo. 1957. Epidemiología del mal del pinta, pinto o carate. *Boletín de la Oficina Sanitaria Panamericana (OSP)* 42 (6): 571.

———. 1963. Reacciones de tipo luetico en el mal del pinto o carate: Historia natural de la enfermedad. *Boletín de la Oficina Sanitaria Panamericana (OSP)* 55 (4): 394.

von Krafft-Ebing, Richard. 1898. *Arbeiten aus dem Gesammtgebiet der Psychiatrie und Neuropathologie*. Leipzig: JA Barth.

Waugh, M.A. 1982. Role played by Italy in the history of syphilis. *British Medical Journal* 58 (2): 92.

12

Threading Worlds Together

Sixteenth-century Taino and Nahuatl, and the seventeenth-century Callinago, some of Amerind peoples of the Encounter display many of the features of Amerindian perspectivism, as well as an inclination towards empiricism. Like some Amerind peoples today, they could acknowledge animals, plants, rocks, and other beings as persons, as beings with agency and subjectivity. These peoples valued perceivable experiences, including those of ecstatic states and dreams. They also experimented with the bodies of different beings to ascertain their composition, their workings, their ways of decaying, and the effects of their substances.

12.1 *Oúpoyem*: Interpenetrating Material Subjectivities Constituting Bodies

The Taino and Callinago understood their bodies and those of many other beings to be dynamic, malleable, and permeable, constituted by and hosting a wide variety of material entities, which could survive after the bodies' death, persisting in its remains, returning to roam in

C. E. Giraldo Herrera, *Microbes and Other Shamanic Beings*,
https://doi.org/10.1007/978-3-319-71318-2_12

atmospheric, geologic, or hydric formations, or becoming embodied into other animals, plants, or rocks. The Taino and the Callinago denominated the general category of these beings as *oúpoyem or acansáncou* and those which were more tightly associated with them, with their crops, as zemes or *çemeijnou* or *ichéiricou.*

These beings had their own whims, intentions, and their own agendas. They required and preferred specific substances for their nourishment and were susceptible to other substances and to other *oúpoyem* which could predate upon them. These entities matured, had genders, reproduced, and aged. Zemes would recognize their kin, had social structures and hierarchies, and acted and feasted in a concerted manner. They guarded certain places, protected, promoted the growth and reproduction of certain persons: humans, animals, plants, rocks belonging to them or sharing their own ethnic affiliations. *Zemes* could empower their hosts affording them specific affects and capacities. They affected the behaviour of their hosts, communicating with or through them.

Zemes or *ichéiricou* were explicitly related with fermentation as is attested by their participation in various brewing techniques, which were understood as their jumping. They were also related with putrefaction processes, as is evidenced by the consultation of corpses as zemes and by the Urayoan experiment. They were pathogenic, inflicting specific diseases and mishaps on strangers trespassing into their places and on enemies or unwarranted predators of their protégées. Through these diseases, they could also punish their protégées for failing to provide nourishment for them or for not following their advice.

The Taino and the Callinago would recognize *oúpoyem* by their aromas. Furthermore, they interacted with these beings through multisensorial experiences induced by fasting and by various substances. Some of these substances they ingested, others they applied to their skin as body paintings, and still others, they applied subcutaneously after cutting and slashing their skin. They would summon these beings, singing and dancing their songs. They would see and hear these beings descending from the sky, singing and dancing to the beat of their heart.

Although some of the substances employed by the Callinago could have been hallucinogenics, many others clearly were not. Many of these substances were extracted from particular rocks, from specific tissues taken from plants, various bodily substances drawn from specific organs of certain animals, or from enemies. Many of these substances were kept in bags and employed as personal adornments, amulets, or charms, like the images of *oúpoyem* carved or engraved in fine woods like guaiacum, seashells, stones, and metal alloys. These artefacts and substances contained the *oúpoyem* that constituted these beings, or means of nourishment or control of *oúpoyem*. These artefacts and substances allowed Callinago to embody or to control their *oúpoyem*, acquiring affects and capacities such as bravery. Although many of these qualities are subjective, Europeans recognized their medical effectiveness.

These artefacts, and the substances composing them, allowed them to imitate, seduce, or mesmerize *oúpoyem*, prey animals, and adversaries, mimicking their movements, their sounds, their ways of being perceived, and thereby also their modes of perceiving the world. The *weapons*, instrument-organs that afford some of the shamanic affects and capacities, are beings like the *oúpoyem*, beings that constitute the material interiority of the bodies and their organs. It was this 'interiority', with its physiological and perceptual implications that the Taino and the Callinago sought to embody—not a metaphysical interiority but a plethora of interpenetrating material subjectivities, partially shared through physical contact, consumption, exposition, activated by signs, which reached beyond humans.

12.2 *Buhities* and *Boyaicou*: Hosts of Zemes, Zemes Themselves

The Taino and the Callinago identified certain hosts of zemes, as zemes themselves, or with words related to *oúpoyem*, such as *buhitie* and *boyáicou*. As peers of zemes, they were able to communicate effectively and interpret their will and advice. They acted as medical practitioners and would engage in intimate interaction with the patient. They would cohabit with the patient, sharing the same diet regime, mimicking their expressions, exposing themselves through close contact (e.g. sucking). In this way, they were able to access similar subjective perspectives, through which they were able to interpret the relation between the pathogenic *oúpoyem* and the patient, further developing an intersubjective relation with the latter. In these circumstances, their zemes would instruct them into suitable remedies, and would assist in the cure of diseases, combating and predating upon enemy zemes or *oúpoyem*.

Practices such as sucking or blowing the disease could have symbolic efficacy as was assumed by missionaries and anthropologists like Lévi-Strauss. However, such explanation is at odds with the claims of shamans. Alternatively, it could be suggested that shamans could indeed be extracting or neutralizing pathogenic substances or organisms through their saliva or through objects previously placed in their mouth.

The *boyáicou* could also employ the affects and capacities afforded by these beings to manage the environment, give advice on matters of war and local politics, and to attack their enemies. The *boyáicou* would sing the songs they had heard *zemes* singing, would reproduce their sounds, and act and narrate their continuously unfolding adventures, what Europeans understood to be their myths.

12.3 The Zemes of the Sun and the Moon, Syphilis, and Other Treponematoses

One of these songs was the *Legend of the Sun and the Moon*, a myth widely spread through the Americas, which in the early Taino and the Nahualt versions explicitly referred to beings that caused syphilis and other treponemal diseases. The Taíno version contains various references to various substances employed to deal with syphilis, such as *Genipa americana* which features in the myth like a zealous guardian and was applied as body painting and employed as a protection against treponemal diseases. The story also mentions the quest for various medicinal herbs employed in baths and purges, *Strombus* spp. seashells, used in plasters to heal treponemal wounds, and metal alloys.

The myth describes a dynamic in which the people hide from the enslaving Sun while attempting to manage the disease; simultaneously they are being fooled into slavery and presumably infected by one of their own. Guahayona, the trickster, had acquired *guaynara* or syphilis through interethnic sexual relations. The disease changed his identity, and he was rewarded with means for controlling the zeme and its pathologies: adornments made of *guanin* or *calloucouli*, a highly prized copper-gold-silver alloy, which European metallurgists were unable to reproduce, and *cibas* or *iámourourou* (seashell). The metal alloy is further identified *híguaili guanin*, the son of Guahayona. In light of a Callinago version, the metal can be identified as the Sun zeme, revealing Guahayona to have become the Moon zeme.

The children who survive Guahayona's deceit become the Cararacol, zemes causing pinta. Their names correspond to Callinago denominations of metal alloy above, suggesting that they were the Sun zeme. Their disease protects and empowers Caracaracol, allowing them to steal women from other communities and to struggle the control of manioc and of the sea from their *oupóyem* Yaya, grandfather of these zemes, causal agent of yaws, and owner of this crop and realm. In this way, the myth explains how the Caracaracol zemes became cultural heroes, rulers, or rather ruling dynamics of these societies.

Various versions of the legend of the Sun and the Moon describe the origin of a community, frequently also their dominion over agriculture or fishing, metallurgy, or pottery, which are associated with a disease. Although later myths do not disclose the name of the specific plague caused by these beings, or confuse it with scabies or leper, which was a European disease, the descriptions correspond with treponemal diseases. Moreover, the areas where these myths have been collected are correlated with the distribution findings of pre-Columbian evidence of treponemal diseases in archaeological remains.

There are various Nahuatl versions of the myth of the Sun and the Moon. The one found in the Florentine Codex, can be cross-referenced with records of Nahuatl medicine, also described in the codex.

Nahuatl physicians explicitly distinguished three varieties of Nanahuatl or treponemal diseases, with epidemiological descriptions coinciding with pinta, syphilis, and yaws, which were, respectively, associated with nobility, crafts-people and prostitutes, and traders—sectors of Nahuatl society respectively afflicted by each of these conditions or maladies. It also mentions various treatments, including baths with copper scrapings, presumably arsenical copper– bronze – alloys. These alloys were regarded as 'gods' faeces', while seashells were regarded as 'gods' food'. The former constrained the development of the pathogens while the latter fed them, employing both means the Nahuatl, and other Amerind peoples, negotiated with the pathogens.

The Nahuatl version of the myth of the Sun and the Moon narrates historical events corresponding to the vacuum of power in the transition between the Toltecs and the Aztecs. The legend tells of the competition between two *teotl* or gods, to become the new Sun illuminating the world, Tecuciztecatl, the richest god from the place of seashells, and Nanahuatzin the humblest *teotl,* lord of syphilis. Both make offerings to the sacrificial fire, the former symbolic, the latter flammable. While the lord of seashells cowers before the fire, the lord of syphilis throws himself into it and becomes the Sun, the master of the Nahuatl. The coward follows, becoming the moon with its pale light and tainted face.

Taking into account medical information, the dating attributed to the myth, we can interpret it as the strife between traders importing arsenical bronze and seashells from South America, versus local crafts-people afflicted by the disease, who imported and developed the metallurgical techniques for the production of the alloy. Ultimately, the latter were able to produce alloys with higher arsenical contents, improving the quality of their tools, musical instruments, and adornments and simultaneously providing a better treatment for the disease, becoming the Sun, while traders turned into the Moon.

Like Caribbean peoples, the Nahualt correlated the expression of disease with sloth, and associated successful cures with the practice of crafts. They endorsed regimes of fasting, followed by diets excluding particular foods, and purges, accompanied by frequent baths with herbal medicines, physical and mental exercise, and sexual abstinence. Amongst the Taino, these regimes were also followed by the *buhities* fostering their attunement with the patient and allowing them develop an intersubjective relation with the pathogenic zeme or *oúpoyem*. Through these various means, Amerindian physicians cultivated long-term relationships with pathogens. It is possible that such a relation, along with the 'cultivation' suggested by the myths, and corroborated by later records the cultural transmission of pinta, generated an attenuated pathogen—the caracaracol, or tekpil-nanahuatl, which conveys an immunity from the more damaging treponematoses.

Early recorded Amerindian myths and associated medical knowledge and practices demonstrate an understanding of the treponemal diseases afflicting these populations. Treponematoses appear as the main characters in the early recorded versions of one of the most widespread Amerindian myths. These myths present the pathogens causing these diseases as key actors associated to a series of politic and economic transformations. The myths acknowledged the causal agents of treponemal diseases as subjects with a history and an intentionality, suggesting that negotiation was possible and indeed necessary. Despite recognizing treponematoses as dreadful diseases, they were also associated with thriving creativity, cultivation, luminosity, and leadership, suggesting that they can be channelled towards creative enterprises.

12.4 Missionary Spirituous Translations

Missionaries, like Pané and Breton, were commissioned to travel to the Americas, learn about Amerindian peoples, and convert them to Christianism. They wrote some of the earliest accounts, most of the dictionaries, and grammars to Amerind languages. Their accounts are reliable insofar the long-term purpose of conversion required accurate descriptions of the language, the social organization, traditions, costumes, beliefs, the natural history, and the resources. However, missionary accounts were also political tools, aimed at justifying the enterprise of conversion and ultimately the conquest of Amerindian peoples, the pillage of their resources, and the usurpation of their lands. These documents identified reasons, goals, paths, obstacles, adversaries, and rewards.

Although missionaries like Breton explicitly declare that the Callinago lacked a spiritual language, he and other missionaries identified notions like *ioüánni*, *oúpoyem or acánsancou*, and *ichéiricou* or *çemeijn* as translations for their notions of souls and spirits. They noted that Amerindian notions differed radically from the Christian understandings[1]: for Amerinds, these were not immortal or immaterial entities, but mutable, and perceivable, living, and social beings, with substance. Moreover,

Amerinds liberally attributed a plethora of these entities, not only to humans but also to animals, trees, and even rocks. Notwithstanding, the notion of the soul was an evidence of the humanity of Amerinds, which compelled the Christian duty of their conversion. The missionaries assumed these differences were 'theological errors' on the part of Amerinds, which they would remediate through the indoctrination into the Christian faith, as other pagans in the past.

The salvation of the 'immortal souls' of Amerindian peoples justified, from a Christian point of view, waging wars against them, taking over their lands and subjugating them. Friars, like Pané, complemented their 'ethical justification' with pragmatic appeals to their respective Crowns, arguing that the conversion would ensure the docility and loyalty of converted Amerinds. Military conquest does not guarantee the submission of the 'other'; ultimately, this requires indoctrination, that is, colonizing the other's mind, a task which Christianity has long excelled.

Missionaries and explorers recognized Amerindian shamans as accomplished physicians and powerful authorities, who managed environmental resources and advised chiefs on local issues and on matters of war. However, or perhaps precisely for those reasons, over the following five hundred years, missionaries sought to disenfranchise shamans from their positions of authority, co-opt their roles in society, as well as the 'shrines' and the 'sacred places' of their zemes. They portrayed the conquest and colonization of the Americas as a holy war and identified the zemes, those peers of Amerindian shamans, as the ultimate adversaries: evil spirits, devils, and false gods. Missionaries, like Breton, sought to isolate shamans and to obscure the references of their practices and understandings. Moreover, characterizing their practices as deceitful quackery, mockery of Christian rituals, and as forms of witchcraft, in which shamans employed inebriating substances to allegedly deceive themselves and the others into idolatry and devil's worship, missionaries made shamans and their knowledge targets of civil and inquisitorial prosecution.

Anthropology has its roots in missionary accounts, and inherited their conceptual frameworks. Consequently, anthropologists continued to translate and understand Amerindian notions with terms such as souls and spirits. Furthermore, anthropologists like Tylor and Levi-Bruhl regarded these notions as characteristic of Amerindians and other 'primitive' ways of thinking, which they typified as animistic.

Like the missionaries, Lévy-Bruhl (1965 [1928]) recognized the depth of the differences between Christian notions and the Amerindian understandings they sought to translate. However, by then the status of souls and spirits had changed. Whereas missionaries and colonial officers assumed those entities to be real, anthropologists, regardless of their personal beliefs, were supposed to maintain a critical scepticism towards supernatural entities. Adhering to the scientific humanist–naturalist paradigm, they vanquished such entities from existence, transforming them into conceptual entities limited to a symbolic realm. This mode of explanation questioned and undermined the understanding of reality of Amerindians and other animists. What they regarded as phenomena, things that can be perceived and therefore were real, were transformed into crude explanations of natural phenomena 'they could not really understand', reducing their perceptually grounded notions to 'rudimentary concepts'.

Later, authors like Evans-Pritchard would note that in many instances magical explanations were not antithetical to naturalist reasoning; rather they went beyond and sought ultimate causation, attributing moral or personal responsibilities, motivating natural phenomena.[2] Likewise, Levi-Strauss noted that myths contained a plethora of accurate naturalistic knowledge, acknowledging them as expressions of situated forms of knowledge. Nevertheless, both were still far from fully recognizing these realities, or from equating these understandings of reality with Science, choosing instead to emphasize their value as sophisticated symbolic models of human relations.

Following this line of reasoning, Lévi-Strauss sought to explain the power of shamans through symbolic efficacy. This explanation and the material on which he substantiates it, replicates early missionary accounts, their interpretations of events, and their accusations of quackery against shamans. This is even more paradoxical because symbolic efficacy is the fundament of the Judeo–Christian faith missionaries were proselytizing. It is this faith that is based on the 'Verb'—a faith in which, by the word of a priest, bread and wine are symbolically, or in their understanding 'ontologically', transformed into the flesh and blood of Christ, a faith in which icons have no power in and of themselves, but solely as symbols of Christian virtues, and as such they operate miracles, which natural science cannot explain.

If we attend to the claims of the Taino and the Callinago, the symbolic efficacy of buhities employed diverse physical means: sounds, substances, and beings. The effigies of zemes, the adornments, and their substances were powerful in themselves. Moreover, according to European observers, including the missionaries, many of these means were materially efficient. Finally, the Taino and the Callinago witnessed how *buhities* addressed through symbols, not just credulous or impressionable humans but beings like the *oúpoyem*—beings which they could perceive.

12.5 Entoptic Microscopy

Some of the substances employed by buhities and later shamans are considered hallucinogenic. However, some of the experiences they produce are more than illusions resulting from brain malfunctions, as missionaries from the sixteenth century onwards and modern neuropsychologists have claimed. These substances allow *muskui ukupsina*, the perception of the body within. These substances disinhibit entoptic vision—a form of microscopy, allowing the perception of the images, the shadows, and reflections of the cells constituting and permeating the body, of the cells flowing to the rhythm of the pulse through the veins of the retina: *ioüánni*, *purpa*, or *utupë*.

Amerindian shamanic practices endorse multiple optical and physiological conditions that further optimize entoptic microscopy and similar forms of perception. Depictions of shamanic visions emphasize the characteristic geometry of shadow formation, corroborating their microscopic origin. These means of perception question the claim that through the use of inebriating substances, buhities and boyáicou were deceiving themselves and others.

Given its profuse irrigation, systemic infections affect early and profoundly the retina, entoptic microscopy might allow the perception of some of the pathogens affecting it, constituting a means of perception which allows the coordination of behavioural and immune responses.

Boyáicou and other Callinago could experience zemes and oúpoyem, beings permeating their bodies, pathogens causing treponematoses. Being able to perceive the other is the fundamental prerequisite to develop an intersubjective relation with them. Based on perceptions available to them, they developed useful knowledge about how to treat some of the diseases afflicting them.

Moreover, pathogens and many of the guest microbes of the body produce those same hallucinogenic substances or their close analogues and employ them as their means of communication. Thus, we should wonder whether some of the more symbolically elaborate 'short-circuit illusions' are microbial messages intended for, or at least, eavesdropped and interpreted by the host.

Callinago notions of *oúpoyem*, Ingano *waira*, Yanomami *xapiri*, and the Tukano *he* are partially based on careful observations of phenomena through means that 'Western' physiology regards as forms of microscopy. Phenomenological access to the microscopic world corroborates a partial commensurability of these forms of knowledge. Under these circumstances, should we insist in translating these notions through the Christian theological concept of spirit, as was done by early missionaries but is now void of weight and value, and relegate their knowledge to the realm of superstitious belief?

12.6 Translating Zemes into the *Seminae*, into Germs, into Microbes

Missionaries sought ways to translate Judeo–Christian beliefs into Amerindian languages, to convert Amerinds into European ontologies. However, the role of missionaries was more complex. Following Augustinian doctrine, the Catholic Church recognized the intrinsic value, the power and political authority derived from accurate naturalist explanation. Missionaries trained to learn about the natural history and the natural philosophy of the pagans they were aiming to convert. Through the Middle Ages, the Church had become the repository of knowledge. Missionaries and colonial officials translated and purified Amerindian natural knowledge. They compiled treatises, collected codices and samples, and translated and incorporated Amerindian understandings of astronomy, metallurgy, botany, zoology, pharmacology, and medicine—the knowledge of buhities. Although the Inquisition restricted the publication of this material, it was circulated and highly treasured in European courts. Substances employed by Amerindian shamans nurtured European apothecaries, constituting important trade products, and numerous medicinal plants were cultivated in European botanical gardens. The ports of entrance from the Americas flourished—economically, intellectually, and artistically.

However, those who returned from the Americas, often rich with gold, spices, and knowledge, were quickly stigmatized as *Indianos,* or Indianized, referring to their being contaminated by Amerindian ways. Partly to overcome that stigma, partly to understand the knowledge arriving from the New World, Europeans turned back to the roots of their culture. Interpreting the new through the old, and the old through the new brought about the Renaissance.

The Encounter brought into contact the realities of Europeans and Amerindians; this included pathogens causing contagious diseases. The demographic catastrophe caused by European diseases in the Amerindian populations is widely recognized. Amerindian pathogens did not make European populations collapse. However, *Treponema pallidum* crossed the ocean, with profound implications for Old World realities. Syphilis spread swiftly as an overwhelming pandemic, which, unlike others, became chronic and tightly imbricated with the flourishing West, afflicting all layers of society.

Syphilis baffled European physicians, challenging them technically and conceptually. However, medics like Fracastoro, who described and named syphilis, were in correspondence with chroniclers like Oviedo and make reference to Anghiera's work. They had access to Amerindian knowledge and followed closely Amerindian regimes of treatment, albeit exclusively on patients, endorsing diets, physical and mental exercise, and sexual abstinence. They employed as treatments baths and purges with decoctions of guaiacum wood, imported from the Americas, as well as mercury. Mercury was the substance employed in European metallurgy to produce golden alloys. Advertising the guaiacum and the mercury cures, Fracastoro adapted Amerindian myths translating their characters and motifs into those of European paganism to make the myths familiar to European audiences. Fracastoro also proposed the earliest *Theory of Contagion*, which, like contemporaneous Amerindian descriptions of zemes, understood contagious disease as the result of an attack by *semina* (seeds or germs): living beings that inhabit, feed, grow, and reproduce in the body and which under suitable circumstances proliferate in the environment, in objects, animals, or persons that transmit the plague to other persons.

The understanding of contagious disease as the effect of attacking living agents provided a theoretical articulation justifying naturalistically the quarantine practices that had been endorsed since the fifteenth century. De la Isla, Fracastoro, and later physicians started conceiving contagious diseases as invading enemies to be exterminated, inaugurating our war on germs. Although guaiacum, mercury, Salvarsan, and penicillin made that goal appear within reach at various times, a definite triumph seems evermore evasive. Paradoxically, in parallel to the development of belligerent attitudes, syphilis became closely associated with the most productive and genial periods of a considerable number of important figures in the Western arts. Its epidemic resounded as deeply in the development of Western culture as it had in its Amerindian counterpart.

Despite foreshadowing the posterior developments of *Germ Theory*, Fracastoro's theory of contagion had very limited impact in Western medical thought of his time. Fracastoro himself remained firmly anchored within the framework of the *Humoral Theory of Disease*, and saw his theory as an extension of the latter rather than as a new paradigm. It would take another century before Leewenhook's microscopy substantiated his claims on the existence of such invisible beings proliferating in the body and the environment. About the same time, the analysis of syphiloids (yaws) led Sydenham to break free from the *Humoral theory*, proposing the *Ontological Theory of Disease*, arguing that diseases existed in and off themselves independently of the balance of humours of individual bodies. Still another couple of centuries would have to pass before Pasteur and Koch developed satisfying means to demonstrate the connection between diseases and microbes, substantiating the *Germ Theory*. This recalcitrance evidences that the notion of living, reproducing *semina* or germs, worked against the grain of the European common sense in established medical thought. Confronted with novel phenomena and novel understandings of those phenomena led first to borrow those understandings adapting them to the established paradigm, and later to rediscover them changing the paradigm.

The translation of Amerindian knowledge was partial and selective. The translations of Nahuatl classifications of treponematoses were misinterpreted by physicians. Europeans would not learn to distinguish syphilis from other treponematoses for a couple centuries, and the issue of whether there were various different diseases or one expressing in different manners remained contested until the development of genomic testing. Moreover, the identification of the relation of pinta with syphilis and yaws eluded Western medicine until the twentieth century. Furthermore, Western prejudices obscured the reasons some Amerindian cultures treasured and purposefully transmitted pinta. Until the 1950s, Western medicine did not recognize the highly effective immune response of the body against pinta, and how this disease provided an immunity towards the more damaging treponematoses. Thus, it did not recognize pinta as a potential vaccine against syphilis and yaws and promoted eradication campaigns which might have led to its extinction.

The reinvention process was not limited to *Germ Theory*, but might include the fundaments of modern pharmacology. Besides arsenical bronze, various Nahuatl herbal remedies against syphilis contain organic arsenic compounds as active principles, anticipating the development of Salvarsan by Ehrlich. Moreover, various versions of the myths present *Genipa* body dyes, an antisyphilic, as a guard against the disease but also as a means to identify the culprit. The early work of Ehrlich focused on the usage of dyes to taint specific kinds of bacteria. This practice led him to the notion of a magic bullet, the idea of a toxic substance with an affinity for particular pathogenic bacteria, restricting to noxious effects to the latter. This notion was realized through the development of Salvarsan from organic arsenical dyes which selectively tainted particular pathogenic microbes. If this phenomenon can be observed macroscopically, applying the *Genipa a.* dyes on infected and non-infected wounds, the myths would foreshadow the logic behind the development of the earliest means of selective chemotherapy by Ehrlich.

Amerindian knowledge and its early appropriation by European physicians had a limited influence on European medicine. It was an ontological scaffold, which remained latent as disarticulate and unproven folk theories, which nevertheless could solve the new problems challenging the old paradigms. Through the course of the following centuries interacting with treponematoses and other diseases, Western physicians (not just Europeans) slowly came to revaluate or rediscover this knowledge and reach similar conclusions, reframing and transforming Western medicine and ontology in the process. The West might be Eurocentric but perhaps it is not so Euromorphic.

The West, wherever that is, is highly polyvalent; it is composed of multiple instantiations, each encompassing a wide variety of syncretic practices, ideas, and identities. Hybridization is not simply a matter of the enculturation to which indigenous communities are being submitted, but more crucially the multifolded simultaneous enculturation processes experienced by those multiple instantiations of the West. Thus, ultimately there is a symmetry underlying the relations of West–non-West, West and non-West are everywhere and nowhere, and everywhere they are slightly different (Fig. 12.1).

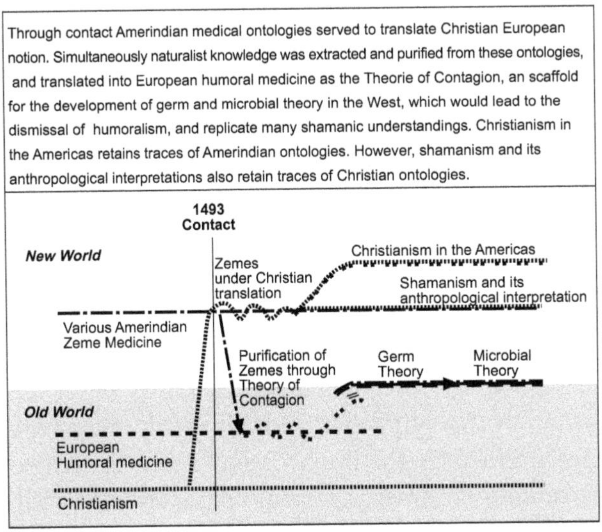

Fig. 12.1 Ontological translations and purifications of knowledge, brought by the Encounter

In the past years, the understanding of microbes has changed further. Contemporary microbes have little in common with the understandings of Pasteur and Koch. These understandings have increasingly converged the descriptions of *oúpoyem* and zemes, or *xapiri.* Like these beings, microbial beings constitute closely knit ecological communities, exchanging signals, and complex information, a variety of *weapons*—means of perceiving and acting upon their world. They can act in concert, cooperating and competing, they also lie for their own benefit. Like *oúpoyem* they are found in rocks, underground, at sea, and in the atmosphere, and are key players in the workings of the environment. They constitute an important proportion of the cells of the body of every plant, animal, and fungi; they are crucial for our functioning, endowing us with many of our specific affects and capacities, promoting our growth and reproduction, affecting moods and behaviours, our perception, cognition, and how we are perceived. Furthermore, some of these beings are zoonotic (causing life-form born) diseases. In other words, they protect their hosts, predating upon their predators, and on strangers to their environments. An issue of increasing concern, given the recent rise of emerging infectious diseases.

Neglecting the conceptual proximity of microbiology and shamanism disregards aspects of shamanic knowledge that have currency and the potential to empower shamans and their communities in relation to biomedicine and environmental sciences. It forecloses the radical transformations of our knowledge of microbiology that may be brought about by taking shamanism into account, at times when alternative approaches to our relations with microbes are called upon.

12.7 Decolonizing Thought, Reclaiming Microbiology

The colonization of thought began with Christian missionary activities. It involved indoctrination, the imposition of Christian religious classificatory frameworks upon Amerindian ontologies, coupled with the appropriation by Europeans of natural knowledge that constituted Amerindian shamanism. The colonizers accompanied these processes with multiple epistemological hostilities and personal attacks aimed at disenfranchising shamans of their political power and disarticulating their forms of knowledge. The decadence of theology and the growth of naturalism in the West emptied theological terms of meaning and credibility, further aggravating the damage previously caused by colonization.

Anthropology has failed to be reflexive regarding its relation towards theology. Preserving missionary translations, anthropology perpetuated, through misattribution, the imposition of Judeo–Christian categories and beliefs upon Amerinds and other animists. Thus, anthropology has been complicit with the continuing colonization of thought. Some of the presumed incommensurabilities between Amerindian and Western ontologies are derived from incommensurabilities intrinsic to Western categories such as religion and science, or humanism and naturalism. The classification of shamanism as a form of religion creates the illusion of its incommensurability with sciences.

Microbiology can provide a better translation for shamanism than anthropology so far has, because it is a result of an interrelated process of syncretism, rather than a purely 'Western' development, as the history of science, STS, and anthropology tend to assume. The similarity between shamanism and microbiology is not the result of analogical developments. That is, microbiology and shamanism are not independent developments, relating through equivalent means to a given reality and reaching similar conclusions. Microbiology is related at its origin with shamanism, it is itself the result of a purification and translation, in the Latourian sense, of Amerindian medical knowledge, which served as an ontological scaffold for its development.

This does not imply reducing shamanism to microbiology and shamanic beings to microbes or vice versa. Quite on the contrary, the reality of the conjugated *microbial-shamanic beings* exceeds either ontology (the shamanic or the microbiological), and because the epistemologies are radically different,[3] there remain many incommensurable aspects. Thus, translation forcibly enlarges the meaning of the terms employed in translation. Both ontologies, microbiology and shamanism, are being extended. However, it is precisely the currently incommensurable aspects that offer the promise of a richer understanding of the world. A common ground also implies that microbiology may have to redraw its understanding of microbes, for instance, considering seriously their subjectivity.

12.8 Remaining Incommensurabilities

The understandings of shamanic and microbial beings have been converging. Nevertheless, these ontologies remain only partially commensurable. As seen in the case of the legend of the Sun and the Moon, microbes might be part of what shamans were describing, but current microbiology does not exhaust shamanic understandings of those beings. Shamans are dealing with a plethora of entities, many of which might be unrelated or only tangentially related to microbes. Furthermore, shamanic epistemologies depend on becoming subjectively involved, privileging and aiming to replicate the perspective of the patient in its relation with the critter, in order to develop an intersubjective relation with the latter.

Moreover, even those beings that do correspond to microbes, for shamans, are persons, and, to make matters more complicated, they are also part of our persons—they are the ones endowing us with some our affects and capacities. In that position, shamanic beings are involved not only in ecologic, physiologic, or emotional dynamics, but also in those that result from social, political, and economic relations.

Meanwhile, microbiology has aimed towards a detached objectifying view of microbes, describing their structure and their mechanisms, reconstructing the conditions of the environments in which they dwell, to address their interrelations and dynamics. Microbiology can provide a highly detailed understanding of some of the general mechanisms and dynamics of the microbial world, in particular of certain model organisms, which allow for multiple inferences. However, the usage of model organisms can lead to misplaced generalizations. Moreover, the objectivity at the base of microbiology, and procedures aimed at insuring the security of its practices, such as destroying cultivated batches after experiments, limit the capacity of microbiologists to relate to their critters individually in the long term, to address their particularities, and how they and we interrelate in vivo beyond experimental circumstances.

Consequently, even if considering microbes as extremely complex beings, and while caring for, admiring, and sometimes even identifying with these critters, most microbiologists are unlikely to acknowledge microbes as persons, endowed with a subjectivity. Moreover, many microbiologists undermine their own claims about the sociality and communication of microbes, taking them as heuristic interpretations.

This reluctance to acknowledge the preconditions and consequences of sociality and communication is partially derived from anthropocentric prejudices associated with human exceptionalism which, as Schaeffer (2005) shows, Western sciences inherited from Christianity. Even while unsubstantiated, these prejudices have remained deeply ingrained in the epistemology of Western sciences and entrenched in disciplinary boundaries.

Trespassing those boundaries, Tsing (2014) has shown how the reproductive relations between fungi radically challenge and expand the possible understandings of kinship. An earnest examination of microbial communication and sociality could bring similar insights. It is in their acknowledgement of subjectivity and in the development of intersubjective relations that shamanic perspectives on microbial-shamanic beings would allow us to earnestly explore their forms of sociality and communication, leading to more enlightening views, relevant not only for the understanding of microbes but for our understanding of our relations with the world. Perhaps then, valuing the patients' experience, we could start to ask what is it that the *Treponema* of a given patient wants? And what can it give in exchange?

However, the incommensurabilities run even deeper and go well beyond microbiology, at least as we understand it at the present. Shamans make many claims that are harder to fathom. For instance, Bacigalupo shows that Mapuche shamans describe shamanic beings, and thereby themselves, as multitemporal beings who transcend temporal barriers in trance and dreams, being and acting simultaneously in the past, the present, and the future. In a similar fashion, Jokic shows how many shamanic practices presuppose action from a distance, either through the 'simultaneous presence' of the shamanic beings in distant locations or through contagious magic, which assumes that substances once in contact remain mutually affected.

I would not pretend that we should simply believe these claims or take them for granted at face value. Shamans are tricky, they themselves acknowledge it, and the interest of anthropologists for the exotic has often lent tricking dynamics a drive of their own. However, we should take them seriously, challenge them and our assumptions, use our imagination, and explore further alliances.[4] Perhaps, then we can become entangled in descriptions that surmount the ontological incommensurability, and really acknowledge their understanding of reality, transcending Western assumptions.

Finally, there are many incommensurabilities, which are more subtle, and though seemingly more mundane are far more pressing. In many areas of low- and middle-income countries, primary health care is largely provided for by traditional forms of medicine, such as shamanism. Biomedicine generally dismisses these forms of knowledge. In the best-case scenarios, it barely tolerates them.

Nevertheless, Jokic[5] shows that biomedicine in the Orinoco, as in other 'peripheries', operates under drastic logistic constrictions and even further epistemological limitations. Biomedical practitioners stationed in these areas often have restricted access to a few Western treatments, some very broad medicines and others highly specific, frequently in line with global agendas. Moreover, these practitioners are frequently students from distant locations, with limited experience in medicine, and even less in the field. With very limited if any skill in local languages, they have to rely on second-language skills of their patients or on translators. These circumstances result in the misplaced concreteness of biomedical knowledge. Without a solid foothold, biomedicine often fails to convince local patients of its procedures and to adequately provide healthcare for them.

The factual knowledge of biomedical practitioners about this environment, its pathogens, and its peoples compares poorly to that of shamans, who combine protracted personal experience with detailed information derived from interactions with their patients, developing a comprehensive understanding of the people, the surrounding ecology, its risks, its diseases, and the available treatments.

On the other hand, globalization is bringing these realities ever closer, exposing peoples like the Yanomami, to multiple diseases with which shamans are not familiar. Simultaneously, with changes in land use in formally forested areas, and the increasing pressure over wildlife and its microorganisms, these critters are opting for more aggressive tactics and finding novel hosts in crops, livestock, and globally interconnected human populations, bringing about the recent rise of contagious diseases. To this day, the biomedical establishment has only been able to react to these threats after they reached epidemic scales. Biomedicine can no longer afford to dismiss shamanism outright—the stakes with emerging and re-emerging infectious diseases are simply too high. Tightly articulating health and environmental management, specifically wildlife management, shamanic practitioners occupy a strategic position for the detection of zoonotic diseases, which in the current globalized context risk becoming emerging infectious diseases. The question is how to translate these forms of knowledge?

Notes

1. See also Viveiros de Castro 2011.
2. Evans-Pritchard 1976.
3. Viveiros de Castro 2004.
4. Lambert et al. 2013.
5. Jokic 2015.

Bibliography

Evans-Pritchard, E.E. 1976. *Witchcraft, oracles and magic among the Azande.* Oxford: Clarendon Press.

Jokic, Zeljko. 2015. *The living ancestors: Shamanism, cosmos and cultural change among the Yanomami of the upper Orinoco.* Oxford: Berghahn.

Lambert, Neill, Yueh-Nan Chen, Yuan-Chung Cheng, Che-Ming Li, Guang-Yin Chen, and Franco Nori. 2013. Quantum biology. *Nature Physics* 9 (1): 10.

Lévy-Bruhl, Lucien. 1965 [1928]. *The soul of the primitive.* Trans. Lilian A. Clare. London: Allen & Unwin.

Schaeffer, Jean-Marie. 2005. La thèse de l'exception humaine. *Communications*: 189–209.

Tsing, Anna Lowenhaupt. 2014. Strathern beyond the Human: Testimony of a spore. *Theory, Culture & Society* 31 (2–3): 221–241. https://doi.org/10.1177/0263276413509114.

Viveiros de Castro, Eduardo. 2004. Exchanging perspectives: The transformation of objects into subjects in Amerindian ontologies. *Common Knowledge* 10 (3): 463–484.

———. 2011. *The inconstancy of the Indian soul: The encounter of catholics and cannibals in 16th-century Brazil.* Trans. Gregory Duff Morton. Chicago: Prickly Paradigm Press.

Bibliography

Abraham, J.J. 1944. The early history of syphilis. *British Journal of Surgery* 32 (126): 225–237.

Adams, D.L., and J.C. Horton. 2002. Shadows cast by retinal blood vessels mapped in primary visual cortex. *Science* 298 (5593): 572.

———. 2003. The representation of retinal blood vessels in primate striate cortex. *Journal of Neuroscience* 23 (14): 5984.

Aghajanian, G.K., and G.J. Marek. 1999. Serotonin and hallucinogens. *Neuropsychopharmacology* 21 (S1): 16S–23S.

Albert, B., and G.G. Gomez. 1997. *Yanomami health: An ethnolinguistic manual*. Belem: PR/McT/CNPq.

Alberti, Benjamin, Severin Fowles, Martin Holbraad, Yvonne Marshall, and Christopher Witmore. 2011. "Worlds otherwise": Archaeology, anthropology, and ontological difference. *Current Anthropology* 52 (6): 896–912. https://doi.org/10.1086/662027.

Alcaraz López, Gloria Margarita, and Leopoldo Yagari Tascón. 2003. The conceptualization of shamanic healing amongst the Colombian Embera: A process of sociocultural and physiological communication. *Investigación Educación y Enfermeria* 21 (2): 60–78.

Altman, L.K. 1987. *Who goes first?: The story of self-experimentation in medicine*. Berkeley: University of California Press.

C. E. Giraldo Herrera, *Microbes and Other Shamanic Beings*,
https://doi.org/10.1007/978-3-319-71318-2

Antal, Georg Michael, Sheila A. Lukehart, and André Z. Meheus. 2002. The endemic treponematoses. *Microbes and Infection* 4 (1): 83–94.

Apffel-Marglin, Frédérique. 1987. *Smallpox in two systems of knowledge*. Helsinki: World Institute for Development Economics Research.

Applegate, R.A., A. Bradley, and W.A. van Heuven. 1990a. Entoptic visualization of the retinal vasculature near fixation. *Investigative Ophthalmology and Visual Science* 31 (10): 2088–2098.

———. 1990b. Entoptic visualization of the retinal vasculature near fixation. *Investigative Ophthalmology & Visual Science* 31 (10): 2088–2098.

Århem, Kaj. 1993a. Ecosofia makuna. In *La selva humanizada: ecología alternativa en el trópico húmedo colombiano*, ed. François Correa, 109–126. Bogotá: ICAN/FEN/CEREC.

———. 1993b. Makuna ecosophy. In *The humanized forest: Alternative ecology in the humid Colombian tropic*, ed. François Correa, 109–126. Bogotá: ICAN/FEN/CEREC.

———. 1996. The cosmic food web: Human-nature relatedness in the Northwest Amazon. In *Nature and society: Anthropological perspectives*, ed. Philippe Descola and Gisli Pálsson, 185–204. London: Routledge.

———. 1998. Powers of place: Landscape, territory and local belonging in Northwest Amazonia. In *Locality and belonging*, ed. Nadia Lovell, 75–98. Oxford: Routledge.

Arte_Amazonia. 2015. *Arte Amazonia*. http://www.arte-amazonia.com/amazonia-artists/painting/. Accessed 19 July 2015.

Atchison, D.A., and G. Smith. 2000. *Optics of the human eye*. Oxford: Butterworth-Heinemann Medical.

Bacigalupo, Ana Mariella. 2016. *Thunder shaman: Making history with Mapuche spirits in Chile and Patagonia*. Austin: University of Texas Press.

Bailey, G.A. 2001. *Art on the Jesuit missions in Asia and Latin America, 1542–1773*. Toronto: University of Toronto.

———. 2005. *Art of colonial Latin America*. London: Phaidon.

Baker, Brenda J., George J. Armelagos, Marshall Joseph Becker, Brothwell Don, Andrea Drusini, Marie Clabeaux Geise, Marc A. Kelley, Iwataro Moritoto, Alan G. Morris, George T. Nurse, Mary Lucas Powell, Bruce M. Rothschild, and Shelley R. Saunders. 1988. The origin and antiquity of syphilis: Paleopathological diagnosis and interpretation [and comments and reply]. *Current Anthropology* 29 (5): 703–737. https://doi.org/10.2307/2743609.

Basso, Keith H. 2009. Speaking with names: Language and landscape among the Western Apache. *Cultural Anthropology* 3 (2): 99–130.

Baumler, Andreas J., and Vanessa Sperandio. 2016. Interactions between the microbiota and pathogenic bacteria in the gut. *Nature* 535 (7610): 85–93. https://doi.org/10.1038/nature18849.

Baylor, D. 1996. How photons start vision. *Proceedings of the National Academy of Sciences of the United States of America* 93 (2): 560–565.

Baylor, D.A., T.D. Lamb, and K.W. Yau. 1979. Responses of retinal rods to single photons. *The Journal of Physiology* 288 (1): 613–634.

Baylor, D.A., B.J. Nunn, and J.L. Schnapf. 1984. The photocurrent, noise and spectral sensitivity of rods of the monkey *Macaca fascicularis*. *Journal of Physiology* 357 (1): 575–607.

Ben Jacob, Eshel, Israela Becker, Yoash Shapira, and Herbert Levine. 2004. Bacterial linguistic communication and social intelligence. *Trends in Microbiology* 12 (8): 366–372. https://doi.org/10.1016/j.tim.2004.06.006.

Ben Jacob, Eshel, Donald S. Coffey, and Herbert Levine. 2012. Bacterial survival strategies suggest rethinking cancer cooperativity. *Trends in Microbiology* 20 (9): 403–410. https://doi.org/10.1016/j.tim.2012.06.001.

Bennett, Jane. 2010. *Vibrant matter: A political ecology of things*. Durham: Duke University Press.

Berlanga, M., B.J. Paster, and R. Guerrero. 2010. The taxophysiological paradox: Changes in the intestinal microbiota of the xylophagous cockroach cryptocercus punctulatus depending on the physiological state of the host. *International Microbiology* 12 (4): 227.

Biocca, Ettore. 1945. Estudios etnobiológicos sobre os indios da regiao do alto rio Negro, Amazonas nota II: transmissao ritual e transmissao criminusa de espiroquetose discromica (puru-puru, Pinta, etc) entre os indios do rio Icana. *Separata de Arquivos de Biología* 29 (265).

Bird, A.C., and R.A. Wealer. 1974. On the retinal vasculature of the human fovea. *Experimental Eye Research* 19 (5): 409–417.

Bird-David, Nurit. 1999. "Animism" revisited: Personhood, environment, and relational epistemology [and comments and reply]. *Current Anthropology* 40: 67–91.

Black, R.J., W.P.R. Gibson, and J.W.R. Capper. 2007. Fluctuating hearing loss in West African and West Indian racial groups: Yaws, syphilis or Menière's disease? *The Journal of Laryngology and Otology* 96 (09): 847–855.

Blackburn, T. 1977. Biopsychological aspects of Chumash rock art. *Journal of California Anthropology* 4 (1): 88–94.

Blaser, M.J. 2006. Who are we? Indigenous microbes and the ecology of human diseases. *EMBO Reports* 7 (10): 956.

Blaser, Martin J. 2010. Harnessing the power of the human microbiome. *Proceedings of the National Academy of Sciences* 107 (14): 6125–6126. https://doi.org/10.1073/pnas.1002112107.

———. 2014. *Missing microbes: How the overuse of antibiotics is fueling our modern plagues*. London: Macmillan.

Boissier de Sauvages, François. 1763a. *Nosologia methodica sistens morborum classes, genera et species, Juxtà Sydenhami mentem et Botanicorum ordinem. Tomi tertii pars prima*. Amsterdam: Frères De Tournes; (c) Bibliothèque interuniversitaire de médecine (Paris).

———. 1763b. *Methodical system of disease nosology*, vol. 3. Amsterdam: Frères De Tournes.

Bókkon, István. 2008. Phosphene phenomenon: A new concept. *Biosystems* 92 (2): 168–174.

Bókkon, István, and Ram Lakhan Pandey Vimal. 2009. Retinal phosphenes and discrete dark noises in rods: A new biophysical framework. *Journal of Photochemistry and Photobiology B: Biology* 96 (3): 255–259.

Borchers, Svenja, Marc Himmelbach, Nikos Logothetis, and Hans-Otto Karnath. 2012. Direct electrical stimulation of human cortex – the gold standard for mapping brain functions? *Nature Reviews Neuroscience* 13 (1): 63–70.

Bouza, Emilio, Rosario Cobo-Soriano, Marta Rodríguez-Créixems, Patricia Muñoz, Marta Suárez-Leoz, and Carlos Cortés. 2000. A prospective search for ocular lesions in hospitalized patients with significant bacteremia. *Clinical Infectious Diseases* 30 (2): 306–312.

Bradley, A., H. Zhang, R.A. Applegate, L.N. Thibos, and A.E. Elsner. 1998. Entoptic image quality of the retinal vasculature. *Vision Research* 38 (17): 2685–2696.

Breitbach, Ulrike B., Michael Niehues, Norberto P. Lopes, Jair E.Q. Faria, and Maria G.L. Brandão. 2013. Amazonian Brazilian medicinal plants described by C.F.P. von Martius in the 19th century. *Journal of Ethnopharmacology* 147 (1): 180–189. https://doi.org/10.1016/j.jep.2013.02.030.

Breton, Raymond. 1665. *Dictionaire caraïbe-françois, Meslé de quantité de Remarques historiques pour l'esclaircissement de la Langue*. Auxerre: Gilles Bovqvet, Imprimeur ordinaire du Roy. Original edition, 1666.

———. 1666. *Dictionaire français-caraïbe*. Auxerre: Gilles Bovqvet, Imprimeur ordinaire du Roy. Original edition, 1666.

Breton, Raymond, and Duna Troiani. 2001. *El caribe insular del siglo XVII Tratado sobre la lengua y la cultura de los Callínago Traducción al español del Dictionnaire caraïbe-français de Raymond Breton*. París: CELIA-CNRS. Original edition, 1665.

Breukel, Tom. 2013. Threepointers on trial: A biographical study of Amerindian ritual artefacts from the pre-Columbian Caribbean. Master's Thesis, Archaeology, University of Leiden.

Bridges, J.H. 1964. *The 'opus majus' of Roger Bacon*, vol. I. Frankfurt: Minerva-Verlag. Original edition, 1266.

Brindley, Giles S. 1973. Sensory effects of electrical stimulation of the visual and paravisual cortex in man. In *Visual centers in the brain*, 583–594. Berlin: Springer.

Brothwell, Don. 1961. The palaeopathology of early British man: An essay on the problems of diagnosis and analysis. *The Journal of the Royal Anthropological Institute of Great Britain and Ireland* 91 (2): 318–344. https://doi.org/10.2307/2844418.

Burgos, J.D., G. Correal-Urrego, and C. Arregoces. 1994. Treponematosis en restos oseos preceramicos de Colombia. *Revista de la Academia Colombiana de Ciencias Exactas, Físicas y Naturales* 19: 273–281.

Burrows, S.M., W. Elbert, M.G. Lawrence, and U. Pöschl. 2009. Bacteria in the global atmosphere – Part 1: Review and synthesis of literature data for different ecosystems. *Atmospheric Chemistry and Physics* 9 (23): 9263–9280. https://doi.org/10.5194/acp-9-9263-2009.

Caicedo, Alhena. 2013. *La alteridad radical que cura. Los nuevos lugares del chamanismo en Colombia*. Tesis de doctorado. École des Hautes Études en Sciences Sociales, París.[Links].

Callaway, J.C., D.J. McKenna, C.S. Grob, G.S. Brito, L.P. Raymon, R.E. Poland, E.N. Andrade, E.O. Andrade, and D.C. Mash. 1999. Pharmacokinetics of Hoasca alkaloids in healthy humans. *Journal of Ethnopharmacology* 65 (3): 243–256. https://doi.org/10.1016/S0378-8741(98)00168-8.

Callegan, Michelle C., Michael S. Gilmore, Meredith Gregory, Raniyah T. Ramadan, Brandt J. Wiskur, Andrea L. Moyer, Jonathan J. Hunt, and Billy D. Novosad. 2007. Bacterial endophthalmitis: Therapeutic challenges and host-pathogen interactions. *Progress in Retinal and Eye Research* 26 (2): 189–203.

Carmichael, Ann G. 1991. Contagion theory and contagion practice in fifteenth-century milan. *Renaissance Quarterly* 44 (2): 213–256. https://doi.org/10.2307/2862709.

Carrasco, P. 1976. La sociedad mexicana antes de la conquista. In *Historia general de México*, ed. D.C. Villegas, 165–288. Ciudad de México D.F.: Colegio de Mexico.

Carrillo, Ana María. 2013. From badge of pride to cause of stigma: Combatting mal del pinto in Mexico. *Endeavour* 37 (1): 13–20. https://doi.org/10.1016/j.endeavour.2012.11.005.

Casas, Eudoro. 2003. *Practices related to the health-disease process among the Tule-kuna: The case of malaria*. Magister en Salud Colectiva MS Colective Health, Facultad de Enfermeria, Universidad de Antioquía.

Castaneda, Carlos. 1968. *The teachings of Don Juan: A Yaqui way of knowledge*. New York: Simon and Schuster.

Castillo-Juáreza, Israel, Violeta Gonzáleza, Héctor Jaime-Aguilara, Gisela Martíneza, Edelmira Linaresb, Robert Byeb, and Irma Romeroa. 2009. Anti-Helicobacter pylori activity of plants used in Mexican traditional medicine for gastrointestinal disorders. *Journal of Ethnopharmacology* 122 (2): 402–405. https://doi.org/10.1016/j.jep.2008.12.021.

Centurion-Lara, A., B.J. Molini, C. Godornes, E. Sun, K. Hevner, W.C. Van Voorhis, and S.A. Lukehart. 2006. Molecular differentiation of *Treponema pallidum subspecies*. *Journal of Clinical Microbiology* 44 (9): 3377.

Cervetto, L., G.C. Demontis, and C. Gargini. 2007. Cellular mechanisms underlying the pharmacological induction of phosphenes. *British Journal of Pharmacology* 150 (4): 383–390.

Césaire, Aimé. 2006 [1955]. *Discursos sobre el colonialismo*, vol. 39. Madrid: Ediciones Akal.

Chapin, M. 1976. Muu ikala: Kuna birth ceremony. In *Ritual and symbol in Central America*, ed. Philip Young and James Howe, 57–66. Eugene Oregon: University of Oregon.

Chapin, N. 1983. *Curing among the San Blas Kuna of Panama*. Doctor in philosophy with a major in cultural anthropology, University of Arizona.

Charbonneau, Mark R., Laura V. Blanton, Daniel B. DiGiulio, David A. Relman, Carlito B. Lebrilla, David A. Mills, and Jeffrey I. Gordon. 2016. A microbial perspective of human developmental biology. *Nature* 535 (7610): 48–55. https://doi.org/10.1038/nature18845.

Christner, Brent C., Cindy E. Morris, Christine M. Foreman, Rongman Cai, and David C. Sands. 2008. Ubiquity of biological ice nucleators in snowfall. *Science* 319 (5867): 1214–1214. https://doi.org/10.1126/science.1149757.

Clark, Hannah-Louise. 2013. Civilization and syphilization: A doctor and his disease in colonial Morocco. *Bulletin of the History of Medicine* 87 (1): 86–114.

Codex Yoalli Ehēcatl/Borgia. 1825–1831. *Codex Yoalli Ehēcatl / Book of Night and Wind / Codex Borgia (Facsimile)*. Edited by British Library: British Library.

———. *Codex Yoalli Ehēcatl / Book of Night and Wind / Codex Borgia. Biblioteca Apostolica Vaticana.*

Coppola, D., and D. Purves. 1996. The extraordinarily rapid disappearance of entopic images. *Proceedings of the National Academy of Sciences of the United States of America* 93 (15): 8001–8004.

Correal, U. 1990. Aguazuque: Evidencias de cazadores, recolectores y plantadores en la altiplanicie de la Cordillera Oriental.

Creighton, John. 1995. Visions of power: Imagery and symbols in late iron age Britain. *Britannia* 26: 285–301.

Cronin, T.W. 2008. Visual ecology. In *The senses: A comprehensive reference*, ed. R. Masland and T.D. Albright, 211–245. New York: Academic Press.

Crosby, Alfred W., Jr. 1969. The early history of syphilis: A reappraisal. *American Anthropologist* 71 (2): 218–227. https://doi.org/10.2307/669854.

Crosby, Alfred W. 2003. *The Columbian exchange: Biological and cultural consequences of 1492*, vol. 2. Westport: Greenwood Publishing Group.

———. 2004. *Ecological imperialism: The biological expansion of Europe, 900–1900.* Cambridge: Cambridge University Press.

Curcio, Christine A., Kenneth R. Sloan, Robert E. Kalina, and Anita E. Hendrickson. 1990. Human photoreceptor topography. *Journal of Comparative Neurology* 292 (4): 497–523.

Cussatlegras, Anne-Sophie, and Patrice Le Gal. 2007. Variability in the bioluminescence response of the dinoflagellate *Pyrocystis lunula. Journal of Experimental Marine Biology and Ecology* 343 (1): 74–81.

David, Lawrence A., Corinne F. Maurice, Rachel N. Carmody, David B. Gootenberg, Julie E. Button, Benjamin E. Wolfe, Alisha V. Ling, A. Sloan Devlin, Yug Varma, and Michael A. Fischbach. 2013. Diet rapidly and reproducibly alters the human gut microbiome. *Nature* 505 (7484): 559–563.

David, G. Biron, Bonhomme Ludovic, Coulon Marianne, and Øverli Øyvind. 2015. Microbiomes, plausible players or not in alteration of host behavior. *Frontiers in Microbiology* 5. https://doi.org/10.3389/fmicb.2014.00775.

Dawkins, R. 1996. *The blind watchmaker: Why the evidence of evolution reveals a universe without design.* New York: WW Norton & Company.

de Avila, Francisco. 2008 [1598]. *Ritos y tradiciones de Huarochirí: manuscrito quechua de comienzos del siglo XVII: versión paleográfica, interpretación fonológica y traducción al castellano*, vol. 12. Lima: Instituto de Estudios Peruanos: Instituto Francés de Estudios Andinos.

de Kievit, Teresa R., and Barbara H. Iglewski. 2000. Bacterial quorum sensing in pathogenic relationships. *Infection and Immunity* 68 (9): 4839–4849. https://doi.org/10.1128/iai.68.9.4839-4849.2000.

de Las Casas, Bartolome. 1951 [1559]. *Historia de las Indias*. Ed. Agustín Millares Carlo. México D.F.: Fondo de Cultura Económica. Original edition, 1559.

de Melo, F. L., J. C. M. de Mello, A. M. Fraga, K. Nunes, and S. Eggers. 2010. Syphilis at the crossroad of phylogenetics and paleopathology. *PLoS Neglected Tropical Diseases* 4 (1):e575(1–11). https://doi.org/10.1371/journal.pntd.0000575.

de Oliveira-Neto, Manoel, Marise S. Mattos, Mauricio A. Perez, Alda Da-Cruz, Octavio Fernandes, João Moreira, Sylvio Gonçalves-Costa, Lúcia R. Brahin, Claudio R. Menezes, and Claude Pirmez. 2000. American Tegumentary Leishmaniasis (ATL) in Rio de Janeiro state, Brazil: Main clinical and epidemiologic characteristics. *International Journal of Dermatology* 39 (7): 506–514.

de Rochefort, Charles. 1665. *The History of the Caribby-Islands: Viz. Barbados, St Christophers, St Vincents, Martinico, Dominico, Barbouthos, Monserrat, Mevis, Antego, &c. in All XXVIII. In Two Books. The first containing the natural; the second, the moral history of those islands. Illustrated with several pieces of sculpture, representing the most considerable rarities therein described. With a Caribbian-vocabulary*. Trans. John Davies. London: T. Dring and J. Starkey.

de Rochefort, Charles, Charles-César comte de Rochefort, César de Rochefort, Jean Baptiste Du Tertre, Louis de Poincy, Raymond Breton, and John Davies. 1965 [1665]. *The history of the Caribby-Islands: Viz. Barbados, St Christophers, St Vincents, Martinico, Dominico, Barbouthos, Monserrat, Mevis, Antego, & c. in all XXVIII. In two books. The first containing the natural; the second, the moral history of those islands. Illustrated with several pieces of sculpture, representing the most considerable rarities therein described. With a Caribbian-vocabulary*: JM [ie John Macocke?] for.

Delaplace, Grégory, and Rebecca Empson. 2007. The little human and the daughter- in-law: Invisibles as seen through the eyes of different kinds of people. *Inner Asia* 9 (2): 197–214.

DeLong, Edward F., Christina M. Preston, Tracy Mincer, Virginia Rich, Steven J. Hallam, Niels-Ulrik Frigaard, Asuncion Martinez, Matthew B. Sullivan, Robert Edwards, and Beltran Rodriguez Brito. 2006. Community genomics among stratified microbial assemblages in the ocean's interior. *Science* 311 (5760): 496–503.

Denton, E.J., and M.H. Pirenne. 1954. The absolute sensitivity and functional stability of the human eye. *Journal of Physiology* 123 (3): 417–442.

Descola, Philippe. 1996. Constructing natures: Symbolic ecology and social practice. In *Nature and society: Anthropological perspectives*, ed. Philippe Descola and Gisli Palsson, 82–101. London: Routledge.

———. 2013. *Beyond nature and culture.* Trans. J. Lloyd. Chicago: University of Chicago Press.

Desowitz, R.S. 1998. *Who gave Pinta to the Santa Maria? Tracking the devastating spread of lethal tropical diseases into America.* London: Harcourt Brace.

Díaz Cíntora, Salvador. 1990. *Xochiquetzal, Seminario de Estudios Prehispánicos para la Descolonización de México.* Ciudad de Mexico D.F.: Universidad Nacional Autónoma de México.

Díaz de Isla, Ruy. 1539. *Tractado co [n] tra el mal serpentino que vulgarmente en España es llamado bubas.* Lisboa: (H)ospital de Todos los Santos.

Dinan, Timothy G., Roman M. Stilling, Catherine Stanton, and John F. Cryan. 2015. Collective unconscious: How gut microbes shape human behavior. *Journal of Psychiatric Research* 63: 1–9. https://doi.org/10.1016/j.jpsychires.2015.02.021.

Douglas, Mary. 1984a. *Purity & danger : An analysis of the concepts of pollution & taboo.* London: Routledge.

———. 1984b. *Purity and danger: An analysis of the concepts of pollution and taboo.* London: Routledge.

Dowson, Thomas A. 1989. Dots and dashes: Cracking the entoptic code in bushman rock paintings. *Goodwin Series* 6: 84–94.

Drew, Rose. 2009. From Saladoid to Taíno: Human behavior from human remains in the Greater Antilles. *Bulletin of the Peabody Museum of Natural History* 50 (1): 167–185.

Du Tertre, Jean-Baptiste. 1954 [1654]. *Histoire générale des isles de S. Christophe, de la Guadeloupe, de la Martinique et autres dans l'Amérique*: chez Jacques Langlois: et Emmanuel Langlois.

Dunny, G.M., T.J. Brickman, and M. Dworkin. 2008. Multicellular behavior in bacteria: Communication, cooperation, competition and cheating. *Bioessays* 30 (4): 296–298.

Durkheim, E. 1969 [1915]. *The elementary forms of the religious life.* Trans. Joseph Ward Swain. New York: Free Press.

Ehrlich, Paul. 1913. Address in pathology, on chemiotherapy. *The British Medical Journal* 2 (2746): 353–359.

Elferink, Jan G. R. 2008. Ethnobotany of the Aztecs. In *Encyclopaedia of the history of science, technology, and medicine in non-western cultures*, ed. Helaine Selin, 826–833. New York: Springer.

Eliade, Mircea. 1976. *El chamanismo y las técnicas arcaicas del éxtasis*. Trans. Ernestina de Champourcin. Bogotá: Fondo de cultura económica.

Evans-Pritchard, E.E. 1976. *Witchcraft, oracles and magic among the Azande*. Oxford: Clarendon Press.

Falkowski, Paul G., Tom Fenchel, and Edward F. Delong. 2008. The microbial engines that drive earth's biogeochemical cycles. *Science* 320 (5879): 1034–1039.

Faust, Franz Xaver. 2001. Indigenous Cauca. In *Possible territories: History, geography and culture of Cauca*, ed. Guido Barona and Cristóbal Gnecco, 259–274. Popayán: Universidad del Cauca.

Fausto, Carlos. 2007. Feasting on people: Eating animals and humans in Amazonia. *Current Anthropology* 48 (4): 497–530. https://doi.org/10.1086/518298.

Feist, Adam M., Markus J. Herrgård, Ines Thiele, Jennie L. Reed, and Bernhard Ø. Palsson. 2009. Reconstruction of biochemical networks in microorganisms. *Nature Reviews Microbiology* 7 (2): 129–143.

Fernández de Oviedo, Gonzalo. 1959 [1535]. Historia general y natural de Indias. In *Biblioteca de Autores Españoles*. Madrid: Biblioteca de Autores Españoles. Original edition, 1935.

Fernandez, E., A. Alfaro, J.M. Tormos, R. Climent, M. Martínez, H. Vilanova, V. Walsh, and A. Pascual-Leone. 2002. Mapping of the human visual cortex using image-guided transcranial magnetic stimulation. *Brain Research Protocols* 10 (2): 115–124.

Flaherty, Gloria. 1989. Goethe and shamanism. *MLN* 104 (3): 580–596. https://doi.org/10.2307/2905047.

———. 1992. *Shamanism and the eighteenth century*. Princeton: Princeton University Press.

Fortis, Paolo. 2010. The birth of design: A Kuna theory of body and personhood. *Journal of the Royal Anthropological Institute* 16 (3): 480–495.

Fracastoro, Hieronimus. 1863 [1535]. *La Syphilis: poema latino/de Gerónimo Fracastor; traducido al castellano e ilustrado con notas por Luis María Ramírez y de las Casas-Leza*. Madrid: Imprenta de de José M. Ducazal.

Fracastoro, Girolamo. 1930 [1546]. *Hieronymi Fracastorii De contagione et contagiosis morbis et eorum curatione: libri III*. Trans. Haven Emerson, *History of medicine series*. New York: Putnam.

Freyre, Gilberto. 1986. *Casa-grande e senzala*, vol. 351: Berkeley: University of California Press.

Gallone, Brigida, Jan Steensels, Troels Prahl, Leah Soriaga, Veerle Saels, Beatriz Herrera-Malaver, Adriaan Merlevede, Miguel Roncoroni, Karin Voordeckers, and Loren Miraglia. 2016. Domestication and divergence of saccharomyces cerevisiae beer yeasts. *Cell* 166 (6): 1397–1410.e16.

Garibay, Angel María. 1961. *Llave del náhuatl: colección de trozos clásicos, con gramática y vocabulario, para utilidad de los principiantes*. Porrua: Ciudad de México D.F. http://mexica.ohui.net/textos/2/.

Gill, S.R., M. Pop, R.T. DeBoy, P.B. Eckburg, P.J. Turnbaugh, B.S. Samuel, J.I. Gordon, D.A. Relman, C.M. Fraser-Liggett, and K.E. Nelson. 2006. Metagenomic analysis of the human distal gut microbiome. *Science* 312 (5778): 1355–1359.

Giraldo Herrera, César Enrique. 2009. *Ecos en el arrullo del mar: Las artes de la marinería en el Pacífico colombiano y sy mimesis en la música y el baile, Prometeo*. Bogotá: Uniandes – Ceso – Departamento de Antropología.

Girtanner, C. 1789. *Abhandlung über die venerische Krankheit*. Gottingen: Johann Christian Dieterich. http://books.google.com/.

Goeje, Claudius Henricus. 1928. *The Arawak language of Guiana*.

———. 1942. De Inwijding tot Medicijnman bij de Arawakken (Guyana) in Tekst en Mythe. *Bijdragen tot de Taal-, Land- en Volkenkunde* 101 (1): 211–276.

Gogarten, J. Peter, and Jeffrey P. Townsend. 2005. Horizontal gene transfer, genome innovation and evolution. *Nature Reviews Microbiology* 3 (9): 679.

Gow, Peter. 1994. River people: Shamanism and history in Western Amazonia. In *Shamanism, history and the state*, ed. Nicholas Thomas and Caroline Humphrey. Ann Arbor: University of Michigan Press.

Grandtner, M.M., and J. Chevrette. 2013. *Dictionary of trees, volume 2: South America: Nomenclature, taxonomy and ecology*. New York: Elsevier Science.

Green, Edward C. 1998. Etiology in human and animal ethnomedicine. *Agriculture and Human Values* 15 (2): 127–131.

———. 1999. *Indigenous theories of contagious disease*. Oxford: Altamira Press.

Green Stócel, Abadio. 2011. *Anmal Gaya Burba: Significados de vida*. Ph.D., Educación, Universidad de Antioquia.

Gross, H., ed. 2008. *Handbook of optical systems vol. 4 survey of optical instruments*. Weinheim: Wiley-VCH Verlag GmbH, Wiley Online Library.

Gruzinski, Serge. 2013. *The mestizo mind: The intellectual dynamics of colonization and globalization*. New York: Routledge.

Gu, Qiang. 2007. *Serotonin involvement in plasticity of the visual cortex. In Monoaminergic modulation of cortical excitability*, 113–124. Boston: Springer.

Gudeman, Stephen, Alberto Rivera, Richard Newbold Adams, Phyllis Deane, Jane I. Guyer, and Lynn Hirschkind. 1989. Colombian conversations: The strength of the earth [and comments and replies]. *Current Anthropology* 30 (3): 267–281.

Guerrero, Ricardo, Carlos Pedrós-Alió, Isabel Esteve, Jordi Mas, David Chase, and Lynn Margulis. 1986. Predatory prokaryotes: Predation and primary consumption evolved in bacteria. *Proceedings of the National Academy of Sciences of the United States of America* 83 (7): 2138–2142.

Guthrie, Stewart. 1995. *Faces in the clouds: A new theory of religion*. New York: Oxford University Press.

Hackett, Cecil John. 1963. On the origin of the human treponematoses: Pinta, yaws, endemic syphilis and venereal syphilis. *Bulletin of the World Health Organization* 29 (1): 7.

Haddock, Steven H.D., Mark A. Moline, and James F. Case. 2010. Bioluminescence in the sea. *Annual Review of Marine Science* 2 (1): 443–493. https://doi.org/10.1146/annurev-marine-120308-081028.

Haley, Bradd J., Arlene Chen, Christopher J. Grim, Philip Clark, Celia Municio Diaz, Elisa Taviani, Nur A. Hasan, Elizabeth Sancomb, Wessam Mahmoud Elnemr, and Muhammad A. Islam. 2012. Vibrio cholerae in a historically cholera-free country. *Environmental Microbiology Reports* 4 (4): 381–389.

Hallett, P.E., F.H.C. Marriott, and F.C. Rodger. 1962. The relationship of visual threshold to retinal position and area. *Journal of Physiology* 160 (2): 364–373.

Hallowell, A.I. 2002 [1960]. Ojibwa ontology, behavior, and world view. In *Readings in Indigenous Religions,* 18–49. London: Continuum.

Hamilton, W.D., and T.M. Lenton. 1998. Spora and Gaia: How microbes fly with their clouds. *Ethology Ecology & Evolution* 10 (1): 1–16.

Haraway, Donna Jeanne. 1985. *A manifesto for cyborgs: Science, technology, and socialist feminism in the 1980s*. San Francisco: Center for Social Research and Education.

———. 2003. *The companion species manifesto: Dogs, people, and significant otherness*. Chicago: Prickly Paradigm Press.

———. 2008. *When species meet*. Minneapolis: University Of Minnesota Press.

———. 2016. *Staying with the trouble: Making Kin in the Chthulucene*. Durham: Duke University Press.

Harper, Kristin N., and George J. Armelagos. 2013. Genomics, the origins of agriculture, and our changing microbe-scape: Time to revisit some old tales and tell some new ones. *American Journal of Physical Anthropology* 152 (S57): 135–152.

Harper, K.N., P.S. Ocampo, B.M. Steiner, R.W. George, M.S. Silverman, S. Bolotin, A. Pillay, N.J. Saunders, and G.J. Armelagos. 2008. On the origin of the treponematoses: A phylogenetic approach. *PLoS Neglected Tropical Diseases* 2 (1): e148.

Harper, Kristin N., Molly K. Zuckerman, Megan L. Harper, John D. Kingston, and George J. Armelagos. 2011. The origin and antiquity of syphilis revisited: An appraisal of Old World pre-Columbian evidence for treponemal infection. *American Journal of Physical Anthropology* 146 (S53): 99–133. https://doi.org/10.1002/ajpa.21613.

Harrison, M. 2013. *Contagion*. New Haven: Yale University Press.

Hastings, J. Woodland. 1996. Chemistries and colors of bioluminescent reactions: A review. *Gene* 173 (1): 5–11.

Hastings, J.W., and E.P. Greenberg. 1999. Quorum sensing: The explanation of a curious phenomenon reveals a common characteristic of bacteria. *Journal of Bacteriology* 181 (9): 2667.

Hastings, J.W., and K.H. Nealson. 1977. Bacterial bioluminescence. *Annual Review of Microbiology* 31 (1): 549–595. https://doi.org/10.1146/annurev.mi.31.100177.003001.

Hayans, Guillermo, S. Henry Wassén, and Niels M. Holmer. 1958. *Nia-ikala: Magical chant to cure madness: Text in Kuna, annotated by the Indian Guillermo Hayans with Spanish translations and comments*. Gothenburg: Etnografiska museet.

Hayden, D. 2003. *Pox: Genius, madness, and the mysteries of syphilis*. New York: Basic Books.

Heinken, Almut, Sahoo Swagatika, R.M. Fleming, and Ines Thiele. 2013. Systems-level characterization of a host-microbe metabolic symbiosis in the mammalian gut. *Gut Microbes* 4 (1): 28–40.

Helmreich, Stefan. 2009. *Alien ocean: Anthropological voyages in microbial seas*. Berkeley: University of California Press.

Henderson, Brian, Stephen Poole, and Michael Wilson. 1996. Microbial/host interactions in health and disease: Who controls the cytokine network? *Immunopharmacology* 35 (1): 1–21.

Hernández, C.A. 2006. Historia natural de la leishmaniasis cutánea y mucocutánea. *Biomédica* 26 (1): 10–12.

Herrera Angel, Marta. 2016. *El conquistador conquistado. Awás, Cuayquer y Sindaguas en el Pacífico colombiano, siglos XVI–XVIII*. Bogotá: Ediciones Uniandes.

Hird, Myra J. 2009. *The origins of sociable life evolution after science studies*. New York: Palgrave Macmillan.

Hodgson, Derek. 2000. Shamanism, phosphenes, and early art: An alternative synthesis. *Current Anthropology* 41 (5): 866–873. https://doi.org/10.1086/317415.

Hofmann, Albert, and Richard Evans Schultes. 1993. *Plantas de los dioses: orígenes del uso de los alucinógenos*. México: FCE.

Holbraad, Martin, Morten Axel Pedersen, and Eduardo Viveiros de Castro. 2014. The politics of ontology: Anthropological positions. *Cultural Anthropology Online*.

Hosler, Dorothy. 1995. Sound, color and meaning in the metallurgy of ancient West Mexico. *World Archaeology* 27 (1): 100–115.

———. 1999. Recent insights into the metallurgical technologies of ancient Mesoamerica. *JOM Journal of the Minerals, Metals and Materials Society* 51 (5): 11–14.

Hsu, Elisabeth. 2000. Spirit (shen), styles of knowing, and authority in contemporary Chinese medicine. *Culture, Medicine and Psychiatry* 24 (2): 197–229.

Hudson, Ellis Herndon. 1958. The treponematoses – or treponematosis? *British Journal of Venereal Diseases* 34 (1): 22.

———. 1961. Endemic syphilis—heir of the syphiloids. *Archives of Internal Medicine* 108 (1): 1.

Hudson, Margaret M., and Robert S. Morton. 1996. Fracastoro and syphilis: 500 years on. *The Lancet* 348 (9040): 1495–1496.

Hughes, Michael F., Barbara D. Beck, Chen Yu, Ari S. Lewis, and David J. Thomas. 2011. Arsenic exposure and toxicology: A historical perspective. *Toxicological Sciences* 123 (2): 305–332. https://doi.org/10.1093/toxsci/kfr184.

Humphrey, Caroline. 2007. Inside and outside the mirror: Mongolian Shamans' mirrors as instruments of perspectivism. *Inner Asia* 9: 173–197.

Husseneder, C. 2010. Symbiosis in subterranean termites: A review of insights from molecular studies. *Environmental Entomology* 39 (2): 378–388.

Idrovo, A.J. 2005. Possible health effects associated with pre-Columbian metallurgy. *Biomédica* 25: 295–303.

Ingold, Tim. 2000. *The perception of the environment: Essays on livelihood, dwelling and skill.* New York: Routledge.

———. 2006. Rethinking the animate, re-animating thought. *Ethnos* 71 (1): 9–20.

———. 2011. *Being alive: Essays on movement, knowledge and description.* Oxford: Routledge.

Ingram, Mrill. 2011. Fermentation, rot, and other human-microbial performances. In *Knowing nature: Conversations at the intersection of political ecology and science studies*, ed. M.J. Goldman, P. Nadasdy, and M.D. Turner, 99–112. Chicago: University of Chicago Press.

Isacsson, Sven Erik. 1993. *Transformations of eternity*. Göteborg: Department of Social Anthropology, University of Göteborg.

James, A.G., J. Casey, D. Hyliands, and G. Mycock. 2004. Fatty acid metabolism by cutaneous bacteria and its role in axillary malodour. *World Journal of Microbiology and Biotechnology* 20 (8): 787–793.

Jara, Fabiola. 1996. La miel y el aguijón. Taxonomia zoológica y etnobiología como elementos en la definición de las nociones de genero entre los andoke (Amazonia colombiana). *Journal de la Société des Américanistes*: 209–258.

———. 2002. The meaning of nominal animal categories among the Caribs of the Guianas. *Anthropos* 97 (1): 117–126. https://doi.org/10.2307/40465620.

Jenner, Edward. 1801. *An inquiry into the causes of the variola vaccine, a disease discovered in some of the western counties of England*. London: D.N. Shuby.

Jokic, Zeljko. 2015. *The living ancestors: Shamanism, cosmos and cultural change among the Yanomami of the upper Orinoco*. Oxford: Berghahn.

Joyce, Thomas Athol. 1916. *Central American and West Indian archaeology: Being an introduction to the archaeology of the states of Nicaragua, Costa Rica, Panama and the West Indies, Central American archaeology*. New York: G. P. Putnam's Sons.

Kanai, Ryota, Leila Chaieb, Andrea Antal, Vincent Walsh, and Walter Paulus. 2008. Frequency-dependent electrical stimulation of the visual cortex. *Current Biology* 18 (23): 1839–1843.

Kehoe, A.B. 2000. *Shamans and religion: An anthropological exploration in critical thinking*. Long Grove: Waveland.

Klüver, Heinrich. 1926. Mescal visions and eidetic vision. *The American Journal of Psychology* 37 (4): 502–515.

———. 1942. Mechanisms of hallucinations. In *Studies in personality*, 175–207. New York: McGraw-Hill.

Knoll, M., J. Kugler, O. Hofer, and S.D. Lawder. 1963. Effects of chemical stimulation of electrically-induced phosphenes on their bandwidth, shape, number and intensity. *Confinia Neurologica* 23 (3): 201–226.

Kohler, U. 1997. El fuego entre los aztecas. In *El fuego. Mitos, ritos y realidades: coloquio internacional Granada , 1–3 de febrero de 1995*, ed. José Antonio González Alcantud and María Jesús Buxó Rey, 199. Barcelona: Anthropos.

Kohn, Eduardo. 2007. How dogs dream: Amazonian natures and the politics of transspecies engagement. *American Ethnologist* 34 (1): 3–24.

———. 2013. *How forests think: Toward an anthropology beyond the human*. Berkeley: University of California Press.

———. 2015. Anthropology of ontologies. *Annual Review of Anthropology* 44: 311–327.

Kopenawa, Davi, and Bruce Albert. 2013. *The falling sky: Words of a Yanomami Shaman*. Trans. Nicholas Elliott and Alison Dundy. Cambridge, MA: Harvard University Press.

Kuby, Janis. 1997. *Immunology*. New York: WH Freeman and Company.

Labat, Jean-Baptiste. 1724. *Nouveau voyage aux isles de l'Amérique contenant l'histoire naturelle de ces pays, l'origine, les moeurs, la religion et le gouvernement des habitans anciens et modernes, les guerres et les evénemens singuliers qui y sont arrivez pendant le long séjour que l'auteur y a fait, le commerce et les manufactures qui y sont établies et les moyens de les augmenter...[par le P. Labat]*.

Laborie, Jean-Claude, and Frank Lestringant. 2006. *Histoire d'André Thevet Angoumoisin, cosmographe du roy, de deux voyages par luy faits aux Indes australes, et occidentales*. Paris: Librairie Droz.

LaFond, R.E., and S.A. Lukehart. 2006. Biological basis for syphilis. *Clinical Microbiology Reviews* 19 (1): 29.

Lambert, Neill, Yueh-Nan Chen, Yuan-Chung Cheng, Che-Ming Li, Guang-Yin Chen, and Franco Nori. 2013. Quantum biology. *Nature Physics* 9 (1): 10.

Langdon, Esther Jean. 2007. Shamans and Shamanisms: Refections on anthropological dilemmas of modernity. *Virtual Brazilian Anthropology* 4 (2): 27–48.

Latour, Bruno. 1987. *Science in action: How to follow scientists and engineers through society*. Cambridge, MA: Harvard University Press.

———. 1988a. *The Pasteurization of France followed by irreductions*. Trans. A. Sheridan and I. Law. Cambridge, MA: Harvard University Press. Original edition, 1984.

———. 1988b. The politics of explanation: An alternative. *Knowledge and Reflexivity: New Frontiers in the Sociology of Knowledge* 10: 155–176.

———. 1993. *We have never been modern*. Hertfordshire: Harvester Wheatsheaf.

———. 1999. *Pandora's hope: Essays on the reality of science studies*. Cambridge, MA: Harvard University Press.

Latour, B. 2009. Perspectivism:'Type'or 'bomb'? *Anthropology Today* 25 (2): 1–2.

Latour, Bruno. 2010. *On the modern cult of the factish gods*. Trans. Barbara Smith and E Weintraub. Durham: Duke University Press.

———. 2014. On selves, forms, and forces. *HAU: Journal of Ethnographic Theory* 4 (2): 261–266.

Latz, Michael I., Michelle Bovard, Virginia VanDelinder, Enrico Segre, Jim Rohr, and Alex Groisman. 2008. Bioluminescent response of individual dinoflagellate cells to hydrodynamic stress measured with millisecond resolution in a microfluidic device. *Journal of Experimental Biology* 211 (17): 2865–2875.

Lévi-Strauss, Claude. 1958a. *Anthropologie structurale*. Paris: Plon.

———. 1958b. *Structural anthropology*. Paris: Plon.

———. 1970. *The raw and the cooked*. Trans. J. & D. Weightman. London: Harper & Row. Original edition, Le cru et le cuit.

———. 1982. *The way of the masks*. Trans. S. Modelski. London: Jonathan Cape.

Lévi-Strauss, C. 1987. Raza e historia. *Antropología Estructural II. Siglo XXI.* México: 304–340.

Lévi-Strauss, Claude. 1988. *The jealous potter*. Trans. B. Chorier. Chicago: University of Chicago Press.

———. 1994 [1974]. *Antropología Estructural*, vol. I. Barcelona: Altaya.

Lévi-Strauss, C. 1995. The structural study of myth. *New Studies in Aesthetics* 26: 85–96.

Lévy-Bruhl, Lucien. 1926. *How natives think*.

———. 1965 [1928]. *The soul of the primitive*. Trans. Lilian A. Clare. London: Allen & Unwin.

Lewis, Charlton T., and Charles Short. 1879. *A latin dictionary*. Oxford: Clarendon Press. Perseus Digital Library Project. Ed. Gregory R. Crane. Tufts University.

Lewis-Williams, J.D. 2004. Neuropsychology and upper palaeolithic art: Observations on the progress of altered states of consciousness. *Cambridge Archaeological Journal* 14 (01): 107–111. https://doi.org/10.1017/S0959774304000083.

Lewis-Williams, J.D., and Thomas A. Dowson. 1990. On neuropsychology and shamanism in rock art. *Current Anthropology* 31 (1): 77–84.

Lewis-Williams, J.D., T.A. Dowson, Paul G. Bahn, H.G. Bandi, Robert G. Bednarik, John Clegg, Mario Consens, Whitney Davis, Brigitte Delluc, Gilles Delluc, Paul Faulstich, John Halverson, Robert Layton, Colin Martindale, Vil Mirimanov, Christy G. Turner, Joan M. Vastokas, Michael Winkelman, and Wylie Alison. 1988. The signs of all times: Entoptic phenomena in upper palaeolithic art [and comments and reply]. *Current Anthropology* 29 (2): 201–245.

Ley, R.E., C.A. Lozupone, M. Hamady, R. Knight, and J.I. Gordon. 2008. Worlds within worlds: Evolution of the vertebrate gut microbiota. *Nature Reviews Microbiology* 6 (10): 776.

Losonczy, Anne Marie. 2006. *The interetnic weft*. Bogotá: ICAHN, IFEA.

Lozupone, Catherine A., Jesse I. Stombaugh, Jeffrey I. Gordon, Janet K. Jansson, and Rob Knight. 2012. Diversity, stability and resilience of the human gut microbiota. *Nature* 489 (7415): 220–230.

Luna, Luis Eduardo. 2011. Indigenous and mestizo use of ayahuasca: An overview. In *The ethnopharmacology of ayahuasca*, ed. Rafael Guimarães dos Santos, 1–21. Kerala: Transworld Research Network.

Lupo, A. 1991. El sol en Jerusalén. *La Palabra y el Hombre* 80: 185–196.

Lyte, Mark. 2013. Microbial endocrinology in the microbiome-gut-brain axis: How bacterial production and utilization of neurochemicals influence behavior. *PLoS Pathogens* 9 (11): e1003726. https://doi.org/10.1371/journal.ppat.1003726.

Lyte, M., and J.F. Cryan. 2014. *Microbial endocrinology: The microbiota-gut-brain axis in health and disease*. New York: Springer.

MacLean, Hope. 2001. Sacred colors and Shamanic vision among the Huichol Indians of Mexico. *Journal of Anthropological Research* 57 (3): 305–323.

Magnus, Ole. 2006. Reconstruction of the rope for the copy of Skuldelev 2: Rope in the Viking Period. In *Connected by the sea*, ed. L.K. Blue, F.M. Hocker, and A. Englert. Oxford: Oxbow Books.

Maier, Steven F., and Linda R. Watkins. 1998. Cytokines for psychologists: Implications of bidirectional immune-to-brain communication for understanding behavior, mood, and cognition. *Psychological Review* 105 (1): 83–107.

Malinowski, Bronislaw. 1948 [1926]. Myth in primitive psychology. In *Magic, science and religion, and other essays*, 93–143. London: Souvenir Press.

Mansilla, J., B.M. Rothschild, C. Pijoa, and C. Rothschild. 2000. Transitions among treponematoses in ancient Mexico. *Chungará (Arica)* 32: 167–174.

Marcos, J., and P. Norton. 1982. From the Yungas of Chinchaysuyo to Cuzco: The role of La Plata Island in Spondylus trade. In *Social and economic organization of the prehispanic Andes*, ed. D.L. Browman, R.L. Burger, and M.A. Rivera, 7–20. Oxford: B.A.R.

Marg, Elwin, and David Rudiak. 1994. Phosphenes induced by magnetic stimulation over the occipital brain: Description and probable site of stimulation. *Optometry and Vision Science* 71 (5): 301–311.

Margulis, Lynn. 1970. *Origin of eukaryotic cells*. New Haven: Yale University Press.

———. 2004. On syphilis & Nietzsche's madness: Spirochetes awake! *Daedalus* 133 (4): 118–125.

Margulis, L., J.B. Ashen, M. Soló, and R. Guerrero. 1993. Composite, large spirochetes from microbial mats: Spirochete structure review. *Proceedings of the*

National Academy of Sciences of the United States of America 90 (15): 6966–6970.

Margulis, Lynn, Michael Chapman, Ricardo Guerrero, and John Hall. 2006. The last eukaryotic common ancestor (LECA): Acquisition of cytoskeletal motility from aerotolerant spirochetes in the proterozoic eon. *Proceedings of the National Academy of Sciences* 103 (35): 13080–13085. https://doi.org/10.1073/pnas.0604985103.

Marshall, Charles Robershaw. 1935. Entoptic phenomena associated with the retina. *British Medical Journal* 19 (4): 177.

———. 1937. An enquiry into the causes of mescal visions. *The Journal of Neurology and Psychopathology* 1 (68): 289.

Medina, Rafael. 1979. Pinta: Una treponematosis endemica de las Americas. *Boletin de la Oficina Sanitaria Panamericana* 86 (3): 69–79.

Mentz, B. V. 2009. Esclavitud y semiesclavitud en el México Antiguo y en la Nueva España (con énfasis en el siglo XVI). *Studia Historica. Historia Antigua* 25 (0).

Mol, Annemarie. 2002. *The body multiple: Ontology in medical practice*. Durham: Duke University Press.

Montiel-Castro, Augusto J., Rina M. González-Cervantes, Gabriela Bravo-Ruiseco, and Gustavo Pacheco-López. 2013. The microbiota-gut-brain axis: Neurobehavioral correlates, health and sociality. *Frontiers in Integrative Neuroscience* 7. https://doi.org/10.3389/fnint.2013.00070.

Mooney, James. 1902. *Myths of the Cherokee*, vol. 19. Washington, DC: Smithsonian Institution. Bureau of American Ethnology.

Morales, Jorge. 1990a. Fauna. Trabajo y enfermedad entre los cuna. In *La selva humanizada: Ecología alternativa en el trópico húmedo colombiano*, ed. François Correa, 167–187. Bogotá: ICAN/FEN/CEREC.

———. 1990b. Fauna: Work and disease among the Kuna. In *The humanized forest: Alternative ecology in the humid Colombian tropic*, ed. François Correa, 167–187. Bogotá: ICAN/FEN/CEREC.

Morales, Diana Marcela. 1994. Enfermedad, curación y jaibanismo. Concepciones embera sobre las enfermedades más comunes. *Boletin del Instituto Francés de Estudios Andinos* 23 (2): 317–357.

Moreland, J.D. 1968. Threshold measurements of the blue arcs phenomenon. *Vision Research* 8 (8): 1093–1106.

Morton, R.S. 1967. The sibbens of Scotland. *Medical History* 11 (04): 374–380.

———. 1990a. Syphilis in art: An entertainment in four parts. *Genitourinary Medicine* 66 (1–4): 33–40, 112–123, 208–221 & 280–294.

———. 1990b. Syphilis in art: An entertainment in four parts. Part 1. *Genitourinary Medicine* 66 (1): 33–40.

Morton, R.S., and S. Rashid. 2001. "The syphilis enigma": The riddle resolved? *Sexually Transmitted Infections* 77 (5): 322.

Murra, John V. 2002. *El mundo andino: población, medio ambiente y economía*, vol. 24. Fondo Editorial PUCP.

Nadell, C.D., J.B. Xavier, and K.R. Foster. 2009. The sociobiology of biofilms. *FEMS Microbiology Reviews* 33 (1): 206–224.

Narby, Jeremy. 1999. *The cosmic serpent: DNA and the origins of knowledge*. New York: Tarcher; Putnam.

Nealson, K.H., and J. Hastings. 2006. Quorum sensing on a global scale: Massive numbers of bioluminescent bacteria make milky seas. *Applied and Environmental Microbiology* 72 (4): 2295.

Nelson, Edward William. 1900. *The Eskimo About Bering Strait*. Washington, DC: U.S.G.P.O.

Nichols, David E. 2004. Hallucinogens. *Pharmacology & Therapeutics* 101 (2): 131–181. https://doi.org/10.1016/j.pharmthera.2003.11.002.

Nimuendaju, Curt. 1914. Die Sagen von der Erschaffung und Vernichtung der Welt als Grundlagen der Religion der Apapocúva-Guaraní. *Zeitschrift für Ethnologie* 46 (2/3): 284–403.

Niño, Juan Camilo. 2007. *Ooyoriyasa. Cosmología e interpretación onírica entre los ette del norte de Colombia*. Bogotá: Universidad de los Andes.

Norris, S.J., D.L. Cox, and G.M. Weinstock. 2001. Biology of treponema pallidum: Correlation of functional activities with genome sequence data. *Journal of Molecular Microbiology and Biotechnology* 3 (1): 37–62.

Norris, Steven, Bruce Paster, Annette Moter, and Ulf Göbel. 2006. The genus treponema. In *The Prokaryotes*, 211–234.

Norris, Steven., Bruce Paster, Annette Moter, and Ulf Göbel. 2006. The Genus Treponema. In The Prokaryotes, ed. Martin Dworkin, Stanley Falkow, Eugene Rosenberg, Karl-Heinz Schleifer and Erko Stackebrandt, 211-234. New York: Springer.

Nutton, Vivian. 1983. The seeds of disease: An explanation of contagion and infection from the Greeks to the renaissance. *Medical History* 27 (1): 1–34.

Oliver, Jose R. 2009. *Caribbean archaeology and ethnohistory : Caciques and cemi idols: The web spun by Taino rulers between Hispaniola and Puerto Rico*. Tuscaloosa: University Alabama Press.

Olsen, Fred. 1974. *On the trail of the Arawaks*. Norman: University of Oklahoma Press.

Osorio, Edison, Gabriel Jaime Arango, Nora Jiménez, Fernando Alzate, Grace Ruiz, David Gutiérrez, Marco Antonio Paco, Alberto Giménez, and Sara

Robledo. 2007. Antiprotozoal and cytotoxic activities in vitro of Colombian Annonaceae. *Journal of Ethnopharmacology* 111 (3): 630–635.

Overing, Joanna. 2006. The stench of death and the aromas of life: The poetics of ways of knowing and sensory process among Piaroa of the Orinoco Basin. *Tipití: Journal of the Society for the Anthropology of Lowland South America* 4 (1): 2.

Pané, Ramón Fray. 1988a [1498]. Relación de Fray Ramón acerca de las antigüedades de los indios, las cuales, con diligencia, como hombre que sabe la lengua de ellos, las ha recogido por mandato del Almirante. In *Relación acerca de las antigüedades de los indios: El primer tratado escrito en América*, ed. Jose Juan Arrom. Bógota: Siglo XXI Ediciones.

———. 1988b [1500]. Relación de Fray Ramón… In *Relación acerca de las antigüedades de los indios: El primer tratado escrito en América*, ed. Jose Juan Arrom. Bógota: Siglo XXI Ediciones.

Paulsen, Allison C. 1974. The thorny oyster and the voice of God: Spondylus and Strombus in Andean prehistory. *American Antiquity* 39: 597–607.

Paxson, Heather. 2008. Post-pasteurian cultures: The microbiopolitics of raw-milk cheese in the United States. *Cultural Anthropology* 23 (1): 15–47.

Paxson, Heather, and Stefan Helmreich. 2013. The perils and promises of microbial abundance: Novel natures and model ecosystems, from artisanal cheese to alien seas. *Social Studies of Science*: 0306312713505003.

Pedersen, Morten A. 2001. Totemism, animism and North Asian indigenous ontologies. *The Journal of the Royal Anthropological Institute* 7 (3): 411–427.

———. 2007. Multiplicity without myth: Theorizing Darhad perspectivism. *Inner Asia* 9 (2): 311–328.

Perry, Rachel J., Peng Liang, Natasha A. Barry, Gary W. Cline, Dongyan Zhang, Rebecca L. Cardone, Kitt Falk Petersen, Richard G. Kibbey, Andrew L. Goodman, and Gerald I. Shulman. 2016. Acetate mediates a microbiome–brain–β-cell axis to promote metabolic syndrome. *Nature* 534 (7606): 213–217. https://doi.org/10.1038/nature18309. http://www.nature.com/nature/journal/v534/n7606/abs/nature18309.html#supplementary-information.

Piaget, Jean. 2009 [1927]. Children's understanding of causality. *British Journal of Psychology* 100 (S1): 207–224.

Pirenne, M.H. 1953. The absolute sensitivity of the eye and the variation of visual acuity with intensity. *British Medical Bulletin* 9 (1): 61.

Porter, Roy. 2006. What is disease. In *The Cambridge history of medicine*, ed. Roy Porter. Cambridge: Cambridge University Press.

Pöschl, U., S.T. Martin, B. Sinha, Q. Chen, S.S. Gunthe, J.A. Huffman, S. Borrmann, D.K. Farmer, R.M. Garland, G. Helas, J.L. Jimenez, S.M. King, A. Manzi, E. Mikhailov, T. Pauliquevis, M.D. Petters, A.J. Prenni, P. Roldin, D. Rose, J. Schneider, H. Su, S.R. Zorn, P. Artaxo, and M.O. Andreae. 2010. Rainforest aerosols as biogenic nuclei of clouds and precipitation in the Amazon. *Science* 329 (5998): 1513–1516. https://doi.org/10.1126/science.1191056.

Pouchet, F.A. 1859. *Hétérogénie ou traité de la génération spontanée, basé sur de nouvelles oxpériences: Avec 3 planches gravées*. JB Baillière.

Pouchet, F. A. 1865. *L'univers: les infiniment grands et les infiniment petits*. Librairie de L. Hachette et Cie.

Pough, Harvey, John Heiser, and William N. Macfarland. 1996. *Vertebrate life*. 4th ed. Upper Saddle River: Prentice Hall.

Poulin, Robert. 2002. Parasite manipulation of host behaviour. In *Behavioural ecology of parasites*, ed. E.E. Lewis, J.F. Campbell, and M.V.K. Sukhdeo. Cambridge, MA: CABI Publishing.

Powell, J.W. 2012 [1895]. *The exploration of the Colorado River and its canyons*. New York: Dover Publications.

Powell, Mary Lucas, and Della Collins Cook, eds. 2005. *The myth of syphilis: The natural history of treponematosis in North America*. Gainesville: University Press of Florida.

Praet, Istvan. 2009. Shamanism and ritual in South America: An inquiry into Amerindian shape-shifting. *Journal of the Royal Anthropological Institute* 15 (4): 737–754.

Prigogine, Ilya. 1998. *Exploring complexity*. New York: Freeman and Company.

Prigogine, Ilya, and Isabelle Stengers. 1997. *The end of certainity: Time, chaos and the laws of nature*. Nueva York: The Free Press.

Pritchard, A. 1842. *A history of infusoria, living and fossil, arranged according to 'die infusionsthierchen' of CG Ehrenberg*. London: Whittaker & CO.

Purkyne, Jan Evangelista. 1819. *Beiträge zur Kenntniss des Sehens in subjectiver Hinsicht*. Prague: Edlen von Wildenbrunn, European Cultural Heritage Online.

Pury, Sybille de, and Marcella Lewis. 2001. *The language of the Callinago people: Father Breton's dictionnaire caraïbe-français (1665) compared with Garifuna*. Paris: Sup-Infor. http://www.sup-infor.com/ultimes/garifuna_a/garifuna_a-ina.htm.

Raj, Kapil. 2013. Beyond postcolonialism… and postpositivism: Circulation and the global history of science. *Isis* 104 (2): 337–347.

Ramawat, K.G. 2008. *Herbal drugs: Ethnomedicine to modern medicine*. Berlin: Springer Berlin Heidelberg.

Ravetz, Jerome R., and Ziauddin Sardar. 1997. Rethinking science. *Futures* 29 (6): 467–470.

Rayer, Piere. 1835. *Traité théorique et pratique des maladies de la peau*, vol. 3. Paris: Chez J.-B. Baillière.

Reed, C.S. 2004. The codpiece: Social fashion or medical need? *Internal Medicine Journal* 34 (12): 684–686.

Reichel-Dolmatoff, G. 1975. *The shaman and the jaguar. A study of narcotic drugs among the Indians of Colombia*. Philadelphia: Temple University Press.

———. 1996. *The forest within: The world-view of the tukano amazonian indians*. San Francisco: Council Oaks Distribution.

———. 1997. *Rainforest shamans: Essays on the tukano indians of the Northwest Amazon*. Cambridge: Green Books.

Reyes Equiguas, Salvador. 2009. El huauhtli y la cotidianidad nahua. *El mundo indígena desde la perspectiva actual* 3 (18).

Riba, Jordi, Peter Anderer, Adelaida Morte, Gloria Urbano, Francesc Jané, Bernd Saletu, and Manel J. Barbanoj. 2002. Topographic pharmaco-EEG mapping of the effects of the South American psychoactive beverage ayahuasca in healthy volunteers. *British Journal of Clinical Pharmacology* 53 (6): 613–628. https://doi.org/10.1046/j.1365-2125.2002.01609.x.

Rincón, Antonio, and Antonio Peñafiel. 1885 [1595]. *Arte mexicana*. México Oficina tip. de la Secretaría de fomento. http://ia331427.us.archive.org/2/items/artemexicana00rincrich/artemexicana00rincrich.pdf.

Rivero, Juan de. 1883 [1736]. *Historia de las misiones de los Llanos de Casanare y los ríos Orinoco y Meta*. Bogota: Biblioteca de la Presidencia de Colombia. http://www.lablaa.org/blaavirtual/historia/hismi/indice.htm.

Robinson, Courtney J., J.M. Brendan, Bohannan, and Vincent B. Young. 2010. From structure to function: The ecology of host-associated microbial communities. *Microbiology and Molecular Biology Reviews* 74 (3): 453–476. https://doi.org/10.1128/mmbr.00014-10.

Román, Gustavo C., and Lydia N. Román. 1986. Occurrence of congenital, cardiovascular, visceral, neurologic, and neuro-ophthalmologic complications in late yaws: A theme for future research. *Reviews of Infectious Diseases* 8 (5): 760–770.

Ronderos Valderrama, Jorge. 2004. El yagé, los curacas (taitas, payés o chámanes) en el Amazonas y el Alto y Bajo Putumayo. Brujería y magia. *Cultura y droga* 11: 91–113.

Rosen, G. 1944. *The specialization of medicine: With particular reference to ophthalmology*. New York: Froben.

Rossell, C., and M.Á.O. Díaz. 2003. *Las mujeres y sus diosas en los códices prehispánicos de Oaxaca*. Mexico: Centro de Investigaciones y Estudios Superiores en Antropología Social (CIESAS) & Porrúa.

Rouse, Irving. 1993. *The Tainos: Rise and decline of the people who greeted Columbus*. New Haven: Yale University Press.

Russell, Andrew, and Elizabeth Rahman, eds. 2015. *The master plant: Tobacco in lowland South America*. London: Bloomsbury.

Safier, Neil. 2008. *Measuring the new world: Enlightenment science and South America*. Chicago: University of Chicago Press.

Sahagun, Bernardo. 1577. *Historia Universal de las Cosas de la Nueva España: Códice Florentino*. Florence: Medicea Laurenciana Library, World Digital Library. http://www.wdl.org/en/item/10096/.

———. 1975 [1580]. *Florentine codex: General history of the things of New Spain*. Santa Fe: School of American Research and University of Utah.

Sallmann, Jean-Michel, and Serge Gruzinski. 1992. *Visions indiennes, visions baroques: les métissages de l'inconscient*. Presses Universitaires de France-PUF.

Sámano, Juan de. 1985. Relación Sámano. In *Verdadera relación de la conquista del Perú. Edición de Concepción Bravo*, ed. Concepción Bravo. Madrid: Historia 16. Original edition, 1527.

Santos-Granero, F. 1998. Writing history into the landscape: Space, myth, and ritual in contemporary Amazonia. *American Ethnologist* 25 (2): 128–148.

———. 2009. Hybrid bodyscapes: A visual history of Yanesha patterns of cultural change. *Current Anthropology* 50 (4): 477–512.

Savage, D.C. 1977. Microbial ecology of the gastrointestinal tract. *Annual Reviews in Microbiology* 31 (1): 107–133.

Schaeffer, Jean-Marie. 2005. La thèse de l'exception humaine. *Communications*: 189–209.

Schiebinger, Londa L. 2009. *Plants and empire*. Cambridge, MA: Harvard University Press.

Schindler, Helmut, and Franz X. Faust. 2000. Interethnic relations amongst healers in the Colombian Southwestern. *Anthropologica* (18–19): 281–294.

Schultes, Richard Evans. 1957. The identity of the malpighiaceous narcotics of South America. *Botanical Museum Leaflets, Harvard University* 18 (1): 1–56.

Schultes, Richard Evans, Bo Holmstedt, and Jan-Erik Lindgren. 1969. De plantis toxicariis e mundo novo tropicale commentationes III: Phytochemical examination of Spruce's original collection of Banisteriopsis caapi. *Botanical Museum Leaflets, Harvard University* 22 (4): 121–132.

Schwiedrzik, C. M. 2009. Retina or visual cortex? The site of phosphene induction by transcranial alternating current stimulation. *Frontiers in Integrative Neuroscience* 3 (6): 1–2. https://doi.org/10.3389/neuro.07.006.2009.

Sekirov, Inna, Shannon L. Russell, L. Caetano, M. Antunes, and B. Brett Finlay. 2010. Gut microbiota in health and disease. *Physiological Reviews* 90 (3): 859–904. https://doi.org/10.1152/physrev.00045.2009.

Seler, Eduard. 1904. *Codex Borgia, eine altmexikanische Bilderschrift der Bibliothek der Congregatio de Propaganda Fide*. 3. Berlin: Loubat.

Sender, Ron, Shai Fuchs, and Ron Milo. 2016. Revised estimates for the number of human and bacteria cells in the body. *bioRxiv*. https://doi.org/10.1101/036103.

Severi, Carlo. 1981. The Kuna soul. *La Ricerca Folklorica* 4: 69–75.

———. 1987. The invisible path: Ritual representation of suffering in Kuna traditional thought. *Res Anthropology and Aesthetics* 14: 66–85.

———. 1993. Talking about souls: The pragmatic construction of meaning in Kuna ritual. In *Cognitive aspects of religious symbolism*, ed. Pascal Boyer, 165–172. Cambridge: Cambridge University Press.

Shorey, Paul, ed. 1969. *Republic, plato in twelve volumes*, Vols. 5 & 6. Trans. Paul Shorey. Cambride, MA: Harvard University Press; William Heinemann, Perseus Digital Library Project. ed. Gregory R. Crane. Tufts University.

Silverman, Mark E., T.J. Murray, and Charles S. Bryan, eds. 2008. *The quotable osler*. Philadelphia: ACP Press.

Sinclair, S.H., M. Azar-Cavanagh, K.A. Soper, R.F. Tuma, and H.N. Mayrovitz. 1989. Investigation of the source of the blue field entoptic phenomenon. *Investigative Ophthalmology and Visual Science* 30 (4): 668.

Sivasundaram, Sujit. 2010. Focus: Global histories of science. Introduction. *Isis; An International Review Devoted to the History of Science and Its Cultural Influences* 101 (1): 95–97.

Šmajs, David, Steven J. Norris, and George M. Weinstock. 2012. Genetic diversity in *Treponema pallidum*: Implications for pathogenesis, evolution and molecular diagnostics of syphilis and yaws. *Infection, Genetics and Evolution* 12 (2): 191–202.

Smith, J. Lawton, and Barry R. Pesetsky. 1967. The current status of treponema cuniculi. Review of the literature. *The British Journal of Venereal Diseases* 43 (2): 117–127.

Smith, J., N. Lauton, J. David, S. Indgin, C.W. Israel, B.M. Levine, J. Justice Jr., J.A. McCrary 3rd, R. Medina, P. Paez, and E. Santana. 1971. Neuro-ophthalmological study of late yaws and pinta. II. The caracas project. *British Medical Journal* 47 (4): 226.

Song, Se Jin, Christian Lauber, Elizabeth K Costello, Catherine A Lozupone, Gregory Humphrey, Donna Berg-Lyons, J. Gregory Caporaso, Dan Knights, Jose C Clemente, and Sara Nakielny. 2013. Cohabiting family members share microbiota with one another and with their dogs. *Elife* 2. https://doi.org/10.7554/eLife.00458.001.

Spivak, Gayatri Chakravorty. 1988. Can the subaltern speak? In *Can the subaltern speak? Reflections on the history of an idea*, ed. Rosalind Morris, 21–78. New York: Columbia University Press.

Stamm, Lola V. 2010. Global challenge of antibiotic-resistant treponema pallidum. *Antimicrobial Agents and Chemotherapy* 54 (2): 583–589. https://doi.org/10.1128/aac.01095-09.

Stanton, T.B., and E. Canale-Parola. 1980. Treponema bryantii sp. nov., a rumen spirochete that interacts with cellulolytic bacteria. *Archives of Microbiology* 127 (2): 145–156.

Stépanoff, C. 2009a. Devouring perspectives: On cannibal shamans in Siberia. *Inner Asia (Cambridge)* 11: 283–307.

———. 2009b. Devouring perspectives: On cannibal shamans in Siberia. *Inner Asia* 11: 283–307.

Stevens Arroyo, Antonio. 1988. *Cave of the Jagua: The mythological world of the Taínos*. Albuquerque: University of New Mexico Press.

Stockman, Andrew, Toby Candler, and Lindsay T. Sharpe. 2010. Human scotopic sensitivity is regulated postreceptorally by changing the speed of the scotopic response. *Journal of Vision* 10 (2): 1–19. https://doi.org/10.1167/10.2.12.

Sydenham, Thomas. 1848. *The works of Thomas Sydenham. M.D.* Edited by William Alexander Greenhill, R. G. Latham, Charles Adlard, James Adlard and publisher Sydenham Society, *Publications of the Sydenham Society*. London: Printed for the Sydenham Society.

Tachau, K.H. 1988. *Vision and certitude in the age of Ockham: Optics, epistemology, and the foundations of semantics, 1250–1345*. Vol. 22. Leiden: Brill Academic Pub.

Takahashi, Jacqueline A., Cássia R. Pereira, Lúcia P.S. Pimenta, Maria Amélia D. Boaventura, and Luiz G.F.E. Silva. 2006. Antibacterial activity of eight Brazilian annonaceae plants. *Natural Product Research* 20 (1): 21–26. https://doi.org/10.1080/14786410412331280087.

Tannock, Gerald W. 2005. Commentary: Remembrance of microbes past. *International Journal of Epidemiology* 34 (1): 13–15. https://doi.org/10.1093/ije/dyh380.

———. 2007. What immunologists should know about bacterial communities of the human bowel. *Seminars in Immunology* 19 (2): 94–105.

Taussig, M.T. 1987. *Shamanism, colonialism, and the wild man: A study in terror and healing*. Chicago: University Chicago Press.

———. 1993. *Mimesis and alterity: A particular history of the senses*. London: Routledge.

Tayler, Donald. 1997. *The coming of the sun: A prologue to Ika sacred narrative*. Oxford: Oxford University: Pitt Rivers Museum.

Taylor, Douglas. 1952. Tales and legends of the Dominica Caribs. *The Journal of American Folklore* 65 (257): 267–279.

———. 1954. A note on the Arawakan affiliation ofTaino. *International Journal of American Linguistics* 20 (2): 152–154.

Thatcher, R.W. 1969. The search for a vaccine for syphilis. An epidemiological approach. *British Medical Journal* 45 (1): 10.

Thevet, André. 1575. *La cosmographie vniverselle*. Paris: Guillaume Chaudiere.

Thiele, Ines, Neil Swainston, Ronan M.T. Fleming, Andreas Hoppe, Swagatika Sahoo, Maike K. Aurich, Hulda Haraldsdottir, Monica L. Mo, Ottar Rolfsson, and Miranda D. Stobbe. 2013. A community-driven global reconstruction of human metabolism. *Nature Biotechnology* 31 (5): 419–425.

Thomas, Christopher M., and Kaare M. Nielsen. 2005. Mechanisms of, and barriers to, horizontal gene transfer between bacteria. *Nature Reviews Microbiology* 3 (9): 711.

Thompson, J. Eric S. 1949. Canoes and navigation of the Maya and their neighbours. *The Journal of the Royal Anthopological Institute of Great Britain and Ireland* 79 (1/2): 69–78.

———. 1967. Creation myths (part 2). *Estudios de cultura maya* 6: 15–43.

Tramont, E.C. 1995. Syphilis in adults: From Christopher Columbus to Sir Alexander Fleming to AIDS. *Clinical Infectious Diseases* 21 (6): 1361–1369.

Troiani, Duna. 2001. *El caribe insular del siglo XVII Tratado sobre la lengua y la cultura de los Callínago. Traducción al español del Dictionaire caraïbe-français de Raymond Breton*. Trans. Duna Troiani. París: CELIA-CNRS Original edition, 1665.

Tsing, Anna Lowenhaupt. 2014. Strathern beyond the Human: Testimony of a spore. *Theory, Culture & Society* 31 (2–3): 221–241. https://doi.org/10.1177/0263276413509114.

———. 2015. *The mushroom at the end of the world: On the possibility of life in capitalist ruins*. Princeton: Princeton University Press.

Turnbaugh, P.J., R.E. Ley, M. Hamady, C.M. Fraser-Liggett, R. Knight, and J.I. Gordon. 2007. The human microbiome project. *Nature* 449 (7164): 804–810.

Turner, Edith. 2003. The reality of spirits. In *Shamanism: A reader*, ed. G. Harvey. London: Routledge.

Turner, T.B., and D.H. Hollander. 1957. Biology of the Treponematoses. *Monograph Series. World Health Organization* (35): 3.

Tylor, Edward B. 1870. The philosophy of religion among the lower races of mankind. *The Journal of the Ethnological Society of London (1869–1870)* 2 (4): 369–381.

———. 1958 [1871]. *Religion in primitive culture*. New York: Harper & Brothers. Original edition, as Chapters XI–XIX of *Primitive Culture*.

Uribe, Carlos Alberto. 2002a. El yajé como sistema emergente: discusiones y controversias. *Documentos del Ceso* 33.

———. 2002b. *Yage as an emergent system: Discussions and controversies*, vol. 33, *Documentos del Ceso*. Bogota: Ediciones Uniandes.

van Leeuwenhoek, Antoni, and Samuel Hoole(Tr). 1800 [co1676–1723]. *The select works of Antony van Leeuwenhoek, containing his microscopical discoveries in many of the works of nature*. London: Sidney and Nicol.

Varela, Gerardo. 1957. Epidemiología del mal del pinta, pinto o carate. *Boletín de la Oficina Sanitaria Panamericana (OSP)* 42 (6): 571.

———. 1963. Reacciones de tipo luetico en el mal del pinto o carate: Historia natural de la enfermedad. *Boletín de la Oficina Sanitaria Panamericana (OSP)* 55 (4): 394.

Vasco, Luis Guillermo. 1985. *Jaibanas: The true men*. Bogota: Biblioteca Banco Popular.

Vasco Uribe, L.G. 1985. *Jaibanás: los verdaderos hombres*. Bogotá: Biblioteca Banco Popular.

Velazco, Salvador. 1998. La imaginacion historiografica de Fernando de Alva Ixtlilxochitl: Etnicidades emergentes y espacios de enunciacion. *Colonial Latin American Review* 7 (1): 33–58. https://doi.org/10.1080/10609169885007.

Velicer, Gregory J. 2003. Social strife in the microbial world. *Trends in Microbiology* 11 (7): 330–337. https://doi.org/10.1016/S0966-842X(03)00152-5.

Velicer, Gregory J., and Helena Mendes-Soares. 2009. Bacterial predators. *Current Biology* 19 (2): R55–R56.

Vilaça, Aparecida. 2005. Chronically unstable bodies: Reflections on Amazonian corporalities. *Journal of the Royal Anthropological Institute* 11 (3): 445–464. https://doi.org/10.1111/j.1467-9655.2005.00245.x.

Villarreal, Luis P. 2004. Can viruses make us human? *Proceedings of the American Philosophical Society* 148 (3): 296–323.
Virtanen, Pirjo Kristiina. 2015. Fatal substances: Apurinã's dangers, movement and kinship. *Indiana* 32: 85–103.
Visvanathan, Shiv. 1998. A celebration of difference: Science and democracy in India. *Science* 280 (5360): 42–43.
———. 2006. Alternative science. *Theory, Culture & Society* 23 (2–3): 164–169. https://doi.org/10.1177/026327640602300226.
Viveiros de Castro, Eduardo. 1998. Cosmological deixis and Amerindian perspectivism. *Journal of the Royal Anthropological Institute* 4 (3): 469–488.
———. 2002a. *Inconstância da alma selvagem*. São Paulo: Cosac Naify.
———. 2002b. *The inconstancy of the savage soul*. São Paulo: Cosac Naify.
———. 2004. Exchanging perspectives: The transformation of objects into subjects in Amerindian ontologies. *Common Knowledge* 10 (3): 463–484.
———. 2007. The crystal forest: Notes on the ontology of Amazonian spirits. *Inner Asia* 9: 153–172.
———. 2011. *The inconstancy of the Indian soul: The encounter of catholics and cannibals in 16th-century Brazil*. Trans. Gregory Duff Morton. Chicago: Prickly Paradigm Press.
Von Helmholtz, H. 1867. *Handbuch der physiologischen Optik*. Leipzig: Voss.
von Krafft-Ebing, Richard. 1898. *Arbeiten aus dem Gesammtgebiet der Psychiatrie und Neuropathologie*. Leipzig: JA Barth.
Wade, N.J., J. Brozek, and J. Hoskovec. 2001. *Purkinje's vision: The dawning of neuroscience*. Oxford: Taylor & Francis.
Wagner, Roy. 2010. *Coyote anthropology*. Lincoln/London: University of Nebraska Press.
Wallis, Robert J. 2002. The bwili or 'flying tricksters' of malakula: A critical discussion of recent debates on rock art, ethnography and shamanisms. *The Journal of the Royal Anthropological Institute* 8 (4): 735–760.
Walls, Gordon L. 1991. Appendix: Mammals. In *Evolution of the eye and visual system*, ed. J. Cronly-Dillon and R.L. Gregory, 467–483. London: Macmillan.
Washer, Peter. 2010. *Emerging infectious diseases and society*. New York: Palgrave Macmillan.
Waugh, M.A. 1982. Role played by Italy in the history of syphilis. *British Medical Journal* 58 (2): 92.
West, Stuart A., Ashleigh S. Griffin, Andy Gardner, and Stephen P. Diggle. 2006. Social evolution theory for microorganisms. *Nature Review Microbiology* 4 (8): 597–607.

Wiener, Margaret J. 2013. Magic,(colonial) science and science studies. *Social Anthropology* 21 (4): 492–509.

Wier, Andrew M., Luciano Sacchi, Michael F. Dolan, Claudio Bandi, James Macallister, and Lynn Margulis. 2010. Spirochete attachment ultrastructure: Implications for the origin and evolution of cilia. *The Biological Bulletin* 218 (1): 25–35.

Willerslev, Rane. 2004. Not animal, not *not*-animal: Hunting, imitation and empathetic knowledge among the Siberian Yukaghirs. *The Journal of the Royal Anthropological Institute* 10 (3): 629–652.

———. 2007. *Soul hunters: Hunting, animism, and personhood among the Siberian Yukaghirs*. Berkeley: University of California Press.

———. 2011. Frazer strikes back from the armchair: A new search for the animist soul. *Journal of the Royal Anthropological Institute* 17 (3): 504–526.

Wilson, Thérèse, and J. Woodland Hastings. 1998. Bioluminescence. *Annual Review of Cell and Developmental Biology* 14 (1): 197–230. https://doi.org/10.1146/annurev.cellbio.14.1.197.

Wittgenstein, Ludwig. 2017 [1922]. *Tractatus Logico-Philosophicus / Logisch-philosophische Abhandlung, side-by-side edition containing the original German, alongside both the English translations. Version 0.45.* Trans. Ogden and Ramsey / Pears and McGuinness. London: Kegan Paul. Kevin Klement. http://people.umass.edu/klement/tlp/.

Xavier, K.B., and B.L. Bassler. 2003. LuxS quorum sensing: More than just a numbers game. *Current Opinion in Microbiology* 6 (2): 191–197.

Xu, Yun, Sarah Dixon, Richard Brereton, Helena Soini, Milos Novotny, Karlheinz Trebesius, Ingrid Bergmaier, Elisabeth Oberzaucher, Karl Grammer, and Dustin Penn. 2007. Comparison of human axillary odour profiles obtained by gas chromatography/mass spectrometry and skin microbial profiles obtained by denaturing gradient gel electrophoresis using multivariate pattern recognition. *Metabolomics* 3 (4): 427–437.

Yong, Ed. 2016. *I contain multitudes: The microbes within us and a grander view of life.* New York: Random House.

Yong, R.N., C.N. Mulligan, and M. Fukue. 2014. *Sustainable practices in geoenvironmental engineering.* 2nd ed. New York: CRC Press.

Ziobrowski, C., E. Cabanillas, and L.G. Palacios. 1996. Estudio de aleaciones cobre-arsénico. *Boletín del Museo del Oro* 41: 131–143.

Index[1]

[1] Note: Page numbers followed by 'n' refer to notes.

C. E. Giraldo Herrera, *Microbes and Other Shamanic Beings*,
https://doi.org/10.1007/978-3-319-71318-2

B

C

W

X

Y

Z

GPSR Compliance
The European Union's (EU) General Product Safety Regulation (GPSR) is a set of rules that requires consumer products to be safe and our obligations to ensure this.

If you have any concerns about our products, you can contact us on

ProductSafety@springernature.com

In case Publisher is established outside the EU, the EU authorized representative is:

Springer Nature Customer Service Center GmbH
Europaplatz 3
69115 Heidelberg, Germany

www.ingramcontent.com/pod-product-compliance
Ingram Content Group UK Ltd.
Pitfield, Milton Keynes, MK11 3LW, UK
UKHW021510290526
12794UKWH00023B/344
9783030100414